Vue.js 3 Design Patterns and Best Practices

Develop scalable and robust applications with Vite, Pinia, and Vue Router

Pablo David Garaguso

‹packt›

BIRMINGHAM—MUMBAI

Vue.js 3 Design Patterns and Best Practices

Group Product Manager: Pavan Ramchandani
Publishing Product Manager: Jane D'Souza
Senior Content Development Editor: Feza Shaikh
Technical Editor: Saurabh Kadave
Copy Editor: Safis Editing
Project Coordinator: Manthan Patel
Proofreader: Safis Editing
Indexer: Manju Arasan
Production Designer: Ponraj Dhandapani
Marketing Coordinators: Anamika Singh, Namita Velgekar, and Nivedita Pandey

First published: May 2023

Production reference: 1240423

Published by Packt Publishing Ltd.
Livery Place
35 Livery Street
Birmingham
B3 2PB, UK.

ISBN: 978-1-80323-807-4

www.packtpub.com

To Elizabeth, Adele, Gabriel, and Eleonora, my wonderful children, whose love and affection kept me going every day. I truly believe the world is a better place because of them.

– Pablo David Garaguso

Foreword

Does the following situation sound familiar to you?

You are invited by a friend to a dinner party. A delicious meal is served, and everyone enjoys the evening. Something about the meal – the combination of flavors – tastes extra special. You want to recreate the meal for yourself to impress your own family and friends. You almost don't dare to ask, but when you do, your host tells you happily about their "secret" method. However, they are an experienced chef and what is simple for them sounds complex to you. Since you feel unsure where to start, your host assures you they will write down and explain all the steps so you can learn how to recreate the meal. By the end of the evening, you have the information you need to produce something special on your own.

By now, you might know where this is going. You chose this book because you are interested in web development. Looking at the field as a whole may feel overwhelming, and you need a clear explanation of the smaller components so you know where to start. Combining the right ingredients (here, libraries and frameworks) with the guidance of an experienced "chef" will enable you to set off on your own journey.

Luckily, you have found a guide written by a master in his field. In my time with Pablo Garaguso, I've found him to have the "magic sauce" that allows him to come to impressive solutions quickly. He applies his knowledge and creativity to inspire his team members. Moreover, he is a wonderful teacher and willing to share freely. He gives you the background information you need to understand how and why to do things a certain way. By following his advice, you will gain a deeper understanding of web development and not only be able to apply the principles in this book but also start to experiment with them and create your own solutions for future projects.

Pablo provides information on setting up a project, constructing UI components, creating single-page apps and progressive web apps, data flow, multithreading, common UI and UX patterns, testing, source control, and deploying your application. He even includes helpful hints on migrating from Vue 2 to Vue 3.

I hope you will enjoy working through the examples provided as much as I did.

I invite you to join in on this journey of discovery!

Olaf Zander
Development Manager, FamilySearch

Contributors

About the author

Pablo David Garaguso was born in Argentina, and since he was a child, he had a special interest in computers. He programmed his first game at the age of 12 years old, and since then, his interest only kept growing. As an adult, he graduated in computer sciences from CAECE University, Mar del Plata, with a specialty in human-computer interfaces. Later in life, he went back to school to work on his MBA at CEMA University, Buenos Aires, with a specialization in corporate entrepreneurship and management. He worked in South America and Europe for almost 20 years in multiple positions. He discovered Vue early on, but it was not until version 2 that he adopted it for his projects. He currently works as a solutions architect and full-stack engineer in Finland, where he resides with his children. His software solutions are used all around the world in specific areas. In his limited spare time, being a natural storyteller, he is also an author of interactive fiction and novels.

About the reviewers

Fotis Adamakis is an experienced software engineer with a strong background in JavaScript, testing, and accessibility. He is proficient in writing clean, maintainable code for large-scale applications, following best practices and industry standards. Fotis has a particular passion for Vue and has developed a high level of expertise in using it.

Aside from his technical skills, Fotis is also a skilled author and speaker on frontend development topics. He has been invited to speak at various conferences, where he shares his knowledge and insights on the latest trends in the field.

He continually seeks new challenges and opportunities to grow and expand his skills.

Simon Nørlund Eldevig is an experienced frontend engineer who has worked in frontend development for over 7 years. His extensive experience has made him an expert in building scalable and responsive web applications.

Throughout his career, Simon has worked with a variety of clients, ranging from small to large enterprises, helping them to develop elegant and user-friendly interfaces. He takes great pride in staying up to date with the latest trends and technologies in frontend development, which has allowed him to create cutting-edge solutions.

Simon is a true frontend enthusiast, which is why he's also an organizer of Vue.js meetups in his local community.

Tamrat Assefa is a full-stack developer from Addis Ababa, Ethiopia. He has a bachelor's degree in software engineering from Addis Ababa University. After landing small projects for local clients while still in school, Tamrat quickly jumped into the web development scene. He has been developing web applications for more than 7 years now and has a very deep knowledge of Vue.js, his favorite JavaScript frontend library. When he is not building features and squashing bugs, you can find him working on cars or playing his favorite game, *Dota 2*.

Table of Contents

3

Setting Up a Working Project 51

4

User Interface Composition with Components 71

5

Single-Page Applications   97

6

Progressive Web Applications   127

7

Data Flow Management   143

11

Bonus Chapter - UX Patterns 221

Preface

Vue 3 is the latest and most performant iteration of the "progressive framework" to create reactive and responsive user interfaces. The framework itself introduces new concepts and approaches to design patterns that could be uncommon in other libraries and frameworks. By learning the basics of the framework and understanding design principles and patterns from software engineering, this book helps you identify the trade-offs of each approach and build solid applications.

It starts with basic concepts and then moves forward through examples and coding scenarios to build incrementally more complex architectures. You will start with a simple page and end with a multithreading, offline, and installable **Progressive Web Application (PWA)**. The content also explores how to use the new test tools available for Vue 3.

Beyond showing you "how things are done," this book helps you to learn how to "think" and "approach" common problems that design patterns have already found a solution for. Avoiding re-inventing the wheel with every project will save you time and make your software better suited for future changes.

Who this book is for

This book targets Vue developers who care about framework design principles and utilize commonly found design patterns in web application development. Learn to use and configure the new bundler (Vite), Pinia (state management), Router 4, web workers, and other technologies to create performant and solid applications. Prior knowledge of JavaScirpt and basic knowledge of Vue would be beneficial.

What this book covers

Chapter 1, The Vue 3 Framework

What is the Vue 3 progressive framework? This chapter introduces the most important aspects of the framework and other key concepts.

Chapter 2, Software Design Principles and Patterns

Software principles and patterns make up the trademarks of good software architecture. This chapter introduces both, with examples for implementation in JavaScript and Vue 3.

Chapter 3, Setting Up a Working Project

With the necessary introductory concepts in place, this chapter sets up a working project that will be used as the base reference for future projects. It will guide you, step by step, on how to start a project using the right tools.

Chapter 4, User Interface Composition with Components

This chapter introduces the concept of user interfaces and leads you into the implementation of a web application, from the conceptual visual design to the development of components to match it.

Chapter 5, Single-Page Applications

This is a key chapter that introduces the Vue Router to create **single-page** web applications.

Chapter 6, Progressive Web Applications

This chapter builds on top of **SPAs** to create PWAs and introduces the use of tools to evaluate their readiness and performance.

Chapter 7, Data Flow Management

This chapter introduces you to key concepts to design and control the flow of data and information within an application and between components. It introduces Pinia as the official state management framework for Vue 3.

Chapter 8, Multithreading with Web Workers

This chapter focuses on improving the performance of a large-scale application using multithreading with web workers. It also introduces more patterns for an easy-to-implement and maintainable architecture.

Chapter 9, Testing and Source Control

In this chapter, we are introduced to the official testing tools provided by the Vue team, as well as the most widespread version control system: Git. The chapter shows how to create test cases for our standalone JavaScript as well as Vue 3 components.

Chapter 10, Deploying Your Application

This chapter presents the necessary concepts to understand how to publish a Vue 3 application on a live production server and how to secure it with a Let's Encrypt certificate.

Chapter 11, Bonus Chapter - UX Patterns, This bonus chapter expands into the concepts of user interface and user experience patterns, to provide a common language between the developer and designer. It presents the common patterns provided by the HTML 5 standard and other common elements.

Appendix: Migrating from Vue 2 to Vue 3

This appendix provides a guide to changes and migration options for experienced Vue 2 developers.

Final Words

In this final chapter, the author briefly summarizes all the concepts learned in each chapter and encourages you to continue your personal development.

To get the most out of this book

This book assumes that you are familiar with and comfortable with web technologies such as JavaScript, HTML, and CSS. Developers interested in expanding their understanding of design patterns and architecture will get the most out of this book. Students and beginners in the world of web applications can also follow this book by paying careful attention to the code examples and using the provided projects from the GitHub repository.

Software/hardware covered in the book	Operating system requirements
Official Vue 3 ecosystem: • Vue 3 framework • Pinia • Vue Router • Vite • Vitest • Vue Testing Tools	Windows, macOS, or Linux
Node.js (any version + v16 LTS)	Windows, macOS, or Linux
Web servers: NGINX, Apache	Windows or Linux
Visual Studio Code	Windows, macOS, or Linux
Chrome browser	Windows, macOS, or Linux

There are no specific hardware requirements considering modern computers, but it is recommended to have at least the following:

• An Intel or AMD CPU of at least 1 GHz

• 4 GB of RAM (more is better)

• At least 10 GB of available storage (for programs and code)

As a general rule, if your computer can run a modern web browser (Chrome/Chromium, Mozilla Firefox, or Microsoft Edge), then it should meet all the requirements to install and run all the developer tools mentioned in this book.

If you are using the digital version of this book, we advise you to type the code yourself or access the code from the book's GitHub repository (a link is available in the next section). Doing so will help you avoid any potential errors related to the copying and pasting of code.

Download the example code files

You can download the example code files for this book from GitHub at `https://github.com/PacktPublishing/Vue.js-3-Design-Patterns-and-Best-Practices`. If there's an update to the code, it will be updated in the GitHub repository.

We also have other code bundles from our rich catalog of books and videos available at `https://github.com/PacktPublishing/`. Check them out!

Code in Action

The *Code in Action* videos for this book can be viewed at `https://packt.link/FtCMS`

Download the color images

We also provide a PDF file that has color images of the screenshots and diagrams used in this book. You can download it here: `https://packt.link/oronG`.

Conventions used

There are a number of text conventions used throughout this book.

`Code in text`: Indicates code words in text, database table names, folder names, filenames, file extensions, pathnames, dummy URLs, user input, and Twitter handles. Here is an example: "The `main.js` file will import and launch the Vue 3 applications."

A block of code is set as follows:

```
<script setup>
    // Here we write our JavaScript
</scrip>
<template>
    <h1>Hello World! This is pure HTML</h1>
</template>
<style scoped>
    h1{color:purple}
</style>
```

When we wish to draw your attention to a particular part of a code block, the relevant lines or items are set in bold:

```
<script>
export default{
    data(){return {_hello:"Hello World"}}
}
</script>
```

Any command-line input or output is written as follows:

```
$ npm install
```

Bold: Indicates a new term, an important word, or words that you see onscreen. For instance, words in menus or dialog boxes appear in **bold**. Here is an example: "In this case, it is worth mentioning the **Boy Scout Principle**, which is similar but applies in groups."

> **Tips or important notes**
> Appear like this.

Get in touch

Feedback from our readers is always welcome.

General feedback: If you have questions about any aspect of this book, email us at customercare@packtpub.com and mention the book title in the subject of your message.

Errata: Although we have taken every care to ensure the accuracy of our content, mistakes do happen. If you have found a mistake in this book, we would be grateful if you would report this to us. Please visit www.packtpub.com/support/errata and fill in the form.

Piracy: If you come across any illegal copies of our works in any form on the internet, we would be grateful if you would provide us with the location address or website name. Please contact us at copyright@packt.com with a link to the material.

If you are interested in becoming an author: If there is a topic that you have expertise in and you are interested in either writing or contributing to a book, please visit authors.packtpub.com.

Share Your Thoughts

Once you've read *Vue.js 3 Design Patterns and Best Practices*, we'd love to hear your thoughts! Scan the QR code below to go straight to the Amazon review page for this book and share your feedback.

https://packt.link/r/1803238070

Your review is important to us and the tech community and will help us make sure we're delivering excellent quality content.

Download a free PDF copy of this book

Thanks for purchasing this book!

Do you like to read on the go but are unable to carry your print books everywhere?

Is your eBook purchase not compatible with the device of your choice?

Don't worry, now with every Packt book you get a DRM-free PDF version of that book at no cost.

Read anywhere, any place, on any device. Search, copy, and paste code from your favorite technical books directly into your application.

The perks don't stop there, you can get exclusive access to discounts, newsletters, and great free content in your inbox daily

Follow these simple steps to get the benefits:

1. Scan the QR code or visit the link below

https://packt.link/free-ebook/9781803238074

2. Submit your proof of purchase
3. That's it! We'll send your free PDF and other benefits to your email directly

1

The Vue 3 Framework

The world wide web of today has changed by many magnitudes since the early days when the internet was just a collection of linked pages for academic and scientific purposes. As the technology evolved and machines became more powerful, more and more features were added to the earlier protocols, and new techniques and technologies competed until finally, standards were adopted. Extra functionality came in the form of plugins for the browser and embedded content. Java applets, Flash, Macromedia, Quicktime, and other plugins were common. It was with the arrival of HTML5 that most, if not all, of these were gradually replaced by standards.

Today, a clear distinction exists between structure, style, and behavior. **Hyper Text Markup Language (HTML)** defines the structural elements that make up a web page. **Cascading Style Sheets (CSS)** provides rules that modify the appearance of HTML elements, including even animations and transformations. And finally, JavaScript is the programming language that provides behavior and can access and modify both HTML and CSS. So many different capabilities also introduced a high level of *complexity* and incompatibility between browsers. This is where libraries and frameworks were born, at first to solve incompatibility issues and standardize appearance, but soon evolved to include other programming paradigms beyond the simple manipulation of HTML and CSS.

Some of the most popular libraries and frameworks today use the **reactive paradigm**. They cleverly make changes in JavaScript to reflect automatically in the HTML/CSS. Vue 3 is the latest version of the progressive framework, which heavily uses the concept of reactivity. It also implements other paradigms and patterns of software design that allow you to build anything from simple interactions in a static web page to complex applications that can even be installed locally and compete with native desktop applications.

In this book, we will explore the Vue 3 framework, and study different design patterns to help us build first-class applications: from simple web pages to powerful **progressive web applications (PWAs)**. Along the way, we will look at best practices and well-proven patterns in software engineering.

This chapter covers the following topics:

- The progressive framework
- Single-file components
- Different syntax options to write components

By the end of this chapter, you will have a basic understanding of where Vue 3 fits into the JavaScript landscape, and what features it provides. For Vue 2 users, there is an appendix to this book, with changes needed to be aware of when migrating an application. As the book progresses, we will build knowledge on top of these concepts.

The progressive framework

Before we describe what Vue is, we need to make the distinction between the terms *library* and *framework*. These are often used interchangeably, but there is a difference, and a good developer should be aware of this when choosing one or the other to build a web application.

Let's have a look at the definitions of these terms:

- A **library** is a collection of reusable code, in the form of functions, classes, and so on, that have been developed by someone else and can be easily imported into your program. It does not prescribe how and where to use it, but normally, they provide documentation on how to use them. It is up to the programmer to decide when and how to implement them. This concept exists in most development languages, to the point that some of them are completely based on the notion of importing libraries to provide functionality.

- A **framework** also has bundles of classes and functions for your use but prescribes specifications that define how the program runs and should be built, with what architecture, and the conditions where or how your code can be used. The key attribute to consider here is that a framework is inverting the control in the application, so it defines the flow of the program and data. By doing so, it emphasizes structures or standards that the programmer should abide by.

Having separated the concepts, now it raises the question of when to use a library and when to use a framework. Before answering that, let's be clear that there is a huge gray area between these two when building real-life applications. In theory, you could build the same application using either one. As always in software engineering, it is a matter of deciding upon the trade-offs for each approach. So, take what comes next with a pinch of salt; it's not a law written in stone:

- You may want to use a *library* when building small to medium-sized applications, or when in need to add additional functionality to your application (in general, you can use additional libraries inside frameworks). There are also exceptions to the "size" guideline. For example, **React** is a library, but there are huge applications built on top of it, such as Facebook. A trade-off to consider is that using only libraries without a framework will need establishing common approaches and more coordination within a team, so management and direction efforts can grow significantly. On the other hand, a library used inside plain JavaScript programming can offer some important performance improvements and give you considerable flexibility.

- You may want to use a *framework* when you build medium to large-sized applications, when you need a structure to help you coordinate the development, or when you want to have a quick start skipping the "basics" of developing common functionality from scratch. There are frameworks that are built on top of other frameworks, for example, **Nuxt** is built on top of **Vue**. The trade-off to consider is that you are prescribed an architecture model to build the application, which often follows a particular approach and way of thinking. You and your team will have to learn about the framework and its limits and live within those boundaries. There is always the chance that your application may outgrow the framework in the future. At the same time, some of the benefits are as follows: easier coordination of work, considerable gains from a head-start, common problems solved true and tested, focus on situations (think shopping applications versus social media, for example), and much more. Depending on the framework, however, you could be facing some small performance loss by the extra processing it takes or difficulties scaling up. It is up to you to weigh up the trade-offs for each case.

So, what is Vue then? By definition, *Vue is a progressive framework* for building user interfaces. Being progressive means that it has the architectural benefits of a framework, but also the speed and modular advantages of a library, as features and functionality can be incrementally implemented. In practice, this means that it prescribes certain models to build your application, but at the same time, allows you to start small and grow as much as you need. You can even use multiple Vue applications on a single page or take over the entire application. You can even import and use other libraries and frameworks if needed. Quite fancy!

Another fundamental concept in Vue is that of **reactivity**. It refers to the capacity of automatically displaying in the HTML the value or changes made to a variable in JavaScript, but also within your code. This is a big part of the magic offered by Vue.

In traditional programming, once a variable is assigned a value, it holds true until programmatically changed. However, in reactive programming, if a variable's value depends on other variables, then when one of those dependencies changes, it will take the new resulting value. Take, for example, the following simple formula:

```
A = B + C
```

In reactive programming, every time B or C changes value, so will A. As you will see later in this book, this is a very powerful model to build user interfaces. In this example, and to be according to the terminology, A is the dependent, and B and C are the dependencies.

In the coming chapters, we will explore this *progressive* attribute as we build the example applications. But before that, we need to see what Vue 3 offers in its most basic form.

Using Vue in your web application

There are several options to use Vue in your web application, and it largely depends on what your objective is:

- To include a small self-contained application or piece of code on a page, you can directly import Vue and code inside a script tag

- To build a larger application, you will need a build tool that takes your code and *bundles* it for distribution

Notice that I use the word *bundle* and not *compile*, as JavaScript applications are interpreted and executed at runtime on the browser. This will become apparent later on when we introduce the concept of **single-file components**.

Let's briefly see an example of the first case in a very simple HTML page:

```html
<html>
<head>
    <script src="https://unpkg.com/vue@3"></script>
</head>
<body>
    <div id="app">
    {{message}}
    </div>
<script>
    const {createApp} = Vue
    createApp({
        data(){
            return {message:'Hello World!'}
        }
    }).mount("#app")
</script>
</body>
</html>
```

In the head section, we define a `script` tag and import Vue from a free **content delivery network (CDN)**. This creates a global variable, `Vue`, which exposes all the methods and functions of the framework. Inside our body tag, we declare a `div` element with `id="app"`. This defines where our small application will be mounted and what part of the page our Vue framework will control. Notice the content of `div`: `{{message}}`. The double curly brackets define a point where the content will be replaced at runtime by the value of the `message` variable that we define in JavaScript. This is called **interpolation** and is the primary way in which a value (string, number, etc.) can be displayed on the web page.

By the end of `body`, we create a script element with our application. We start by extracting the `createApp` function from Vue and use it to create an application by passing an object. This object has specific fields that define a **component**. In this case, this component only exposes a `data()` method that, in turn, returns an object. The field names in this object will be treated as reactive variables that we can use in our JavaScript as well as in the HTML. Finally, the `createApp()` constructor returns the Vue 3 application instance, so we chain the invocation and call the `mount()` method to, well, mount our humble application to the element with the `app` ID. Notice that we are using CSS selectors as the argument (the pound sign indicates the `id` property, hence `id="app"` is selected by `#app`).

As this method of using Vue is not that common (or popular), we will focus on greater things and will use a **bundler** to organize our workflow and have a significantly better developer experience... but first, we need to know a bit more about Vue and what makes it so great.

The bundler way, a better way...

As you can imagine, importing Vue directly into a web page would only work for very small applications. Instead, Vue is structured in the concept of *components*, which are reusable isolated sets of JavaScript code, HTML, and CSS that behave as a unit. You can think of them as building blocks to compose a web page. Obviously, a browser knows nothing about this so we will use a *bundler* to transform our application into something the browser can interpret, with the added benefit of running a number of optimizations in the process. Here is where the "framework" part comes into action, as it prescribes how these components should be written and what methods need to contain.

When using a bundler, it will wrap up all of our code into one or more JavaScript files that the browser will load at runtime. The execution workflow in a browser for a Vue application could be simplified as follows:

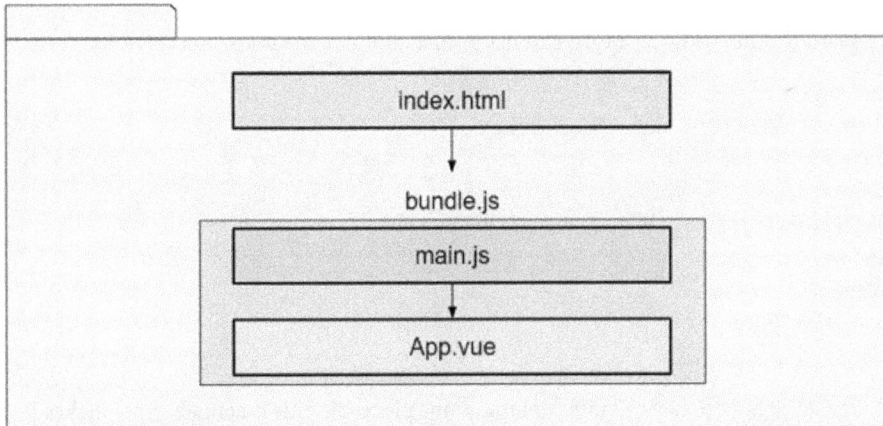

Figure 1.1: A very simplified view of our application execution order when using a bundler

The browser will load the index.html page as usual and then load and execute the bundle.js file, just like any other JavaScript. The bundler will have packaged all our files and execute them in a prescribed order:

1. The main.js file will import and launch the Vue 3 application.
2. Then, it will start the page composition from the *main* component, here encapsulated in the App.vue file. This component will spawn other components, thus forming a tree of components that make up the page.

Don't worry if this sounds a bit strange right now. We will see these concepts in action as we advance building our example applications throughout the book. In *Chapter 3, Setting Up a Working Project*, we will start a simple application using this same diagram.

So far, you have had a view of what libraries and frameworks are, and just a quick peek into what Vue has to offer. What is important to remember is that in the modern JavaScript world, it is common to use bundlers to help us organize our applications and optimize the code for the browser. We will work with the official Vue 3 bundler, **Vite**, later on. But first, we need a few more basic concepts.

Understanding single-file components

As you may have guessed, the App.vue file mentioned before is a **single-file component** (**SFC**), one of the great contributions of Vue. In this type of file, we can describe the HTML, CSS, and JavaScript that defines the component. The web page is then composed in a hierarchy of components, starting from an entry point (traditionally named App.vue) down to the last customized button, if you will. We will discuss components in depth in *Chapter 4, User Interface Composition with Components*, but for now, keep in mind that this is the way to go, as prescribed by the framework. If you have experience in an object-oriented language, this may look familiar (and you wouldn't be wrong).

An SFC is a plain text file with the .vue extension that contains the following sections:

```
<script setup>
    // Here we write our JavaScript
</scrip>
<template>
    <h1>Hello World! This is pure HTML</h1>
</template>
<style scoped>
    h1{color:purple}
</style>
```

It may look strange at first, to have all this content in one place, but this is actually what makes it great. Here is a description of each section:

- A **script** tag that surrounds our JavaScript, and depending on the syntax, it exports an object with well-defined fields. In practice, this becomes a *module*, which is the modern way to split code in JavaScript. Notice also that we are using a modifier attribute, `setup`. This will define the application interface that we are going to use to write our code in Vue. We could also declare the `lang="ts"` attribute to use TypeScript instead of plain JavaScript.

- A **template** tag surrounds the HTML for our component. Here, we can use HTML elements, other components, directives, and so on. A great advantage of Vue is that we can use plain HTML to write our HTML. This may sound obvious, but other libraries handle this completely differently and have their own syntax for it. However, Vue 3 also allows the use of other syntaxes through the use of bundler plugins. We are not left out of options here.

- A **style** tag, where we will place the CSS for our component. In this case, we use the `scoped` attribute, which will encapsulate the rules and limit them to our component, thus preventing them from "bleeding out" to the rest of the application. In the same way as with previous sections, we can also use different syntax to write the styles, as long as it is supported by the bundler.

> **Best practice**
> Always scope your styles, unless you are defining styles on a parent component or CSS variables that you want explicitly to be passed on to the entire application. For application-wide styles, use a separate CSS file.

The important concept to remember is that an SFC contains these three elements that define a single component. The bundler application will do its magic and separate each part and place it where it belongs, so the browser can interpret them properly. We will use the fast and new **Vite** for this in *Chapter 3, Setting Up a Working Project*, and in *Chapter 4, User Interface Composition with Components*, we will delve in-depth into components and how to handle the flow of control and information between them. But first, let's take a look at how we write our components.

Different strokes – options, composition, and script setup API

The classical way to describe a component in Vue 2 has been branded as the *Options API*. To maintain backward compatibility, the same syntax is also supported in Vue 3. However, there is also a new syntax named the *Composition API*, which is what we will use in this book.

The *Options API* is inherited from Vue 2 and prescribes that a component is defined by an object with defined fields, none of which is actually mandatory. Moreover, some of them have defined parameters and expected outputs. For example, these are the most common fields to use (also, a non-exclusive list):

- `data` should be a function that returns an object whose fields would become reactive variables.

- `methods` is an object that contains our functions. These functions have access to the reactive variables from data by using the `this.variableName` format.

- `components` is an object where each field provides a name for the template, and the value points to the constructor of another component (child to the current one).

- `computed` is an object whose attributes define "computed" properties. Each member is then a function or an object that can be used as reactive variables in our template and code. Functions will be read-only, and objects could include logic to read and write values to them. This concept will be clarified as we see code examples in *Chapter 3, Setting Up a Working Project*.

- `props` and `emits` declare parameters to receive data from the parent component and declare events that get dispatched to the parent component. This provides a formal way to communicate and pass data between related components, but is not the only one, as we will see in *Chapter 7, Data Flow Management*.

- Life cycle hooks methods are a series of functions that are triggered during the life cycle of the component.

- Mixins are objects that describe a common functionality that could be shared across multiple components. This is not the only way to reuse code in Vue 3. The use of mixins in the Options API caused some complications that gave birth to the Composition API. We will not deal with mixins in detail, but will see other approaches to share functionality between components (such as "composables").

This syntax is well-defined but has some limitations. For small components, it is too much scaffolding code, and for large components, the code organization suffers greatly and is very verbose. Plus, in order to reference the reactive variables declared in the `data` section or other methods, the internal code has to use the `this` keyword (e.g., `this.data_variable_name` or `this.myMethod()`). The `this` keyword refers to the created instance of the component. The problem is when the reserved word `this` changes meaning depending on the scope and context of use. There are other drawbacks that have appeared over time that led to the creation of the Composition API. However, this syntax is relevant and fully supported by Vue 3. One advantage of this is that you can easily migrate code from Vue 2 (within certain considerations, as shown later in the *Appendix – Migrating from Vue 2*).

The Composition API exposes a method called `Setup()` that is executed before the component is mounted. In this method, we import functions and components, declare variables, and so on, that define our component instead of declaring them as "options." This means that you can write your code in more of a JavaScript way This gives you the freedom to import, reuse, and organize your code better.

Let's see a comparison between the two approaches with a reactive variable, `_hello="Hello World"`:

Options API

```
<script>
 export default{
    data(){return {_hello:"Hello World"}}
 }
</script>
```

Composition API

```
<script>
  import {ref} from "vue"
  export default{
    setup(){
      const _hello=ref("Hello World")
      return {_hello}
    }
  }
</script>
```

In the Options API, we just use the `data` field to return an object whose fields will turn into reactive variables. Vue will take care of interpreting the object. However, notice how, in the Composition API, we need to first import from Vue the `ref` constructor, which will create a reactive constant or variable for us. The end result is the same, but here, we have more fine control over what is done and where. When using the new Vite bundler, this fine control of what gets imported into our components may result in faster code building and development times.

At first sight, it seems that the Composition API is more verbose than the Options API, and so it is for such a trivial example. However, as our component begins to grow, this becomes the opposite. Still, verbose… So, there is an alternative syntax for the Composition API called *script setup*, and is the one we will use in this book. Let's compare now how this component looks with this new syntax:

Composition API – script setup

```
<script setup>
    import {ref} from "vue"
    const _hello=ref("Hello World")
</script>
```

Two lines of code! That is hard to beat. Because we added the `setup` attribute in the `script` tag, the bundler knows everything we do here is in the realm of the Composition API, and all the functions, variables, and constants are automatically exposed to the template. There's no need to define exports. If we need something, we import it directly and use it. Also, we now have a few extra advantages, such as the following:

- We can have reactive and non-reactive variables displayed in our template

- We know that all the code is executed before the component is mounted

- The syntax is closer to vanilla JavaScript (a big plus!!!), so we can organize our code to our convenience and pleasure

- Smaller bundle size (did I mention this before? Yes, it is important!)

But wait, you may notice that I'm defining a reactive variable as a *constant*! Yes, I am! And no, it is not an error. In JavaScript, a constant points to a particular immutable value, which, in this case, is an object, but this applies only to the object, not to its members. The `ref()` constructor returns an object, so the constant applies to the object reference and we *can* change the value of its members. If you have worked with pointers in Java, C, or a similar language, you may recognize this concept as the use of **pointers**. But all this comes at a cost. In order to access and modify the value, now we need to access the `value` attribute from the object. Here is an example:

```
_hello.value="Some other value";
```

But, at the same time, nothing has changed in the way to access this variable in the template:

```
<div>{{_hello}}</div>
```

So, in brief, every time a variable is declared as reactive using the `ref()` constructor, you need to reference its value with the `constant_name.value` format, and just as `constant_name` in the template (HTML). When the constant name is used in the template, Vue already knows how to access the value and you don't need to reference it explicitly as in JavaScript.

> **Tip**
> Adopt a code convention so you'll know when an identifier refers to a variable, constant, function, class, and so on.

Exploring built-in directives in Vue 3

Vue also provides special HTML attributes called **directives**. A directive is declared in the opening tag of an HTML element and will affect or provide dynamic behavior or functionality to that element. We can also create our own directives in Vue. Those provided by the framework have a special notation starting with `v-`. As for the purpose of this book, let's explain the most commonly used Vue directives:

v-bind: (shorthand ":")

The `v-bind:` directive binds the value of an HTML attribute to the value of a JavaScript variable. If the variable is reactive, each time it updates its value, it will be reflected in the html. If the variable is not reactive, it will be used only once during the initial rendering of the HTML. Most often, we use only the `:` shorthand prefix (semi-colon). For example, the `my_profile_picture` reactive variable contains a web address to a picture:

```
<img :src="my_profile_picture">
```

The `src` attribute will receive the value of the `my_profile_picture` variable.

v-show

This directive will show or hide the element, without removing it from the document. It is equivalent to modifying the CSS `display` attribute. It expects a variable that gives a Boolean value (or something that can be interpreted as true or non-empty). For example, the `loading` variable has a Boolean value:

```
<div v-show="loading">...</div>
```

The `div` will appear when the `loading` variable is true.

It is important to keep in mind that `v-show` will use the style of the object to display it or not, but the element will still be part of the **Document Object Model (DOM)**.

v-if, v-else, and v-else-if

These directives behave as you would expect with conditional sentences in JavaScript, showing and hiding the element based on the value resolved by the expression passed. They are similar to `v-show` in the sense that they will show or hide the element, but with the difference that they remove completely the element from the DOM. Because of this, it can be expensive computationally if used improperly at a large scale with elements that switch their state often, as the framework has to perform more operations to manipulate the DOM, as opposed to `v-show`, when only the display style needs to change.

> **Note**
>
> Use `v-if` to show or display elements that will not toggle once shown or hidden (and preferred when the initial state is hidden). Use `v-show` if an element will switch states often. This will improve the performance when displaying large lists of elements.

v-for and :key

These two attributes, when combined, behave like a `for` loop in JavaScript. They will create as many copies of the element as prescribed in the iterator, each one with the corresponding interpolated value. It is extremely useful to display collections of data items. The `:key` attribute is used internally to keep track of changes more efficiently, and must reference a unique attribute of the item being looped on – for example, the `id` field of an object, or the index in an array when the indexes won't change. Here is an example:

```
<span v-for="i in 5" :key="i"> {{i}} </span>
```

This will display five `span` elements on the web page with the interpolation of i showing the following:

```
1  2  3  4  5
```

v-model

This directive is pure magic. When attached to an input element (input, textarea, select, etc.), it will assign the value returned by the HTML element to the referenced variable, thus keeping the DOM and JavaScript state in synchronization – something that is called **two-way binding**. Here is an example:

```
<input type="text" v-model="name">
```

When the user enters text in HTML, the `"name"` variable in JavaScript will immediately have that value assigned. In these examples, we are using primitive data types such as numbers and strings, but we can also use more complex values such as objects or arrays. More of this will come in *Chapter 4, User Interface Composition with Components*, when we see components in depth.

v-on: (and the shorthand @)

This directive behaves a bit differently than the ones seen before. It expects not a variable, but a function or an expression, and it ties an HTML event to a JavaScript function to execute it. The event needs to be declared immediately after the colon. For example, to react to a `click` event on a button, we would write the following:

```
<button v-on:click="printPage()">Print</button>
```

When the button triggers the `click` event, the JavaScript `"printPage()"` function will be executed. Also, the shorthand for this directive is more commonly used, and we will use that from now on in this book: just replace the `v-on:` with `@`. Then, the previous example becomes the following:

```
<button @click="printPage()">Print</button>
```

You can find the complete list of built-in directives in the official documentation here: `https://vuejs.org/api/built-in-directives.html`. We will see others as we move forward.

So far, we have seen that Vue 3 applications are built with components that we can use in our HTML and that we create using SFCs. The framework also provides us with directives to manipulate HTML elements, but that is not all. In the next section, we'll see that the framework also provides some handy prebuilt components for us to use.

Built-in components

The framework also provides us with several built-in components that we can use without explicitly importing them into each SFC. I have provided here a small description of each one, so you can refer to the official documentation for the syntax and examples (see `https://vuejs.org/api/built-in-components.html`):

- `Transition` and `TransitionGroup` are two components that can work together to provide animations and transition to elements and components. They need you to create the CSS animations and transition classes to implement the animation when inserting or removing elements into the page. They are mainly (or often) used when you are displaying a list of elements with `v-for/:key` or `v-if/v-show` directives.

- KeepAlive is another wrapper component (meaning that it surrounds other components) used to preserve the state (internal variables, elements, etc.) when the component wrapped inside is no longer on display. Usually, component instances are cleared out and "garbage collected" when they are unmounted. KeepAlive keeps them cached so their state is restored when they come back on display.

- Teleport is a brand-new component in Vue 3, that allows you to transport the HTML of the component into another location anywhere on the page, even outside the component tree of your application. This helps in some cases where you need to display information outside your component but it has to be processed by your component's internal logic.

- Suspense is a new component in Vue 3, but is still in an experimental phase, so its future is uncertain at the time of this writing. The basic idea is to display "fallback" content until all the asynchronous child components/elements are ready to be rendered. It is provided as a convenience since there are patterns that you could use to solve this problem. We will see some later on.

- Component-is is a special element that will load a component at runtime, as prescribed by the content of a variable – for example, if we need to display a component based on the value of a variable, and the use of other directives may be cumbersome. It can also be used to render HTML elements. Let's see an example:

```
<script setup>
    import EditItem from "EditItem.vue"
    import ViewItem from "ViewItem.vue"
    import {ref} from "vue"
    const action=ref("ViewItem")
</script>
<template>
    <component :is="action"></component>
    <button @click="action='EditItem'">Edit</button>
</template>
```

In this simple example, when the user clicks the Edit button, the action value will change to EditItem, and the component will be swapped in place. You can find the documentation here: https://vuejs.org/api/built-in-special-elements.html.

With the idea of frameworks and components, we are now better prepared to move forward.

Book code conventions

In this book, we will use a set of code conventions and guidelines that are good practices for Vue 3. They will help you not only understand the examples of this book but also the code in the wild that you may come across, as more and more developers use it. Let's start from the beginning.

Variables and props

These are always in lowercase and spaces are replaced with an underscore, for example, `total_count` and `person_id`..

Constants

References to injected objects start with a $ (dollar) sign, for example, `$router`, `$modals`, and `$notifications`.

References to reactive data start with _ and are typed in snake case, for example, `_total` and `_first_input`.

References to constant values are all in capital letters, for example, `OPTION` and `LANGUAGE_CODE`.

Constructor functions for injected dependencies will start with `use`, for example, `const $store=useStore()`.

Classes and component names

These are written in PascalCase (each word starts with an upper case letter), for example, `Person`, `Task`, and `QueueBuilder`.

Functions, methods, events, and filenames

These are written in camel case, for example, `doSubscribe()` and `processQueue()`

Instances

Instances will have the abstract name, followed by the word `Service` in the case of plain JavaScript objects that provide functions, `Model` for state models, and so forth. We will use services to encapsulate functionality.

Here's an example: `const projectService=new ProjectService()`.

> **Tip**
> With your team, always use a code convention that all agree upon. This will make the code more readable and maintainable. It can be also recommended to use a linter (a processor to capture conventions in your code).

As mentioned, these code conventions are gaining popularity, so you may see them in multiple projects. However, these are not mandatory standards and most definitely are not prescribed by the framework. You can write all in capital letters if that is your style, but what really matters is that you and your team define and abide by your own conventions in a consistent manner. What matters in the end, is that we all have a common language when writing code.

Summary

This chapter has gone from the basics of libraries and frameworks to Vue 3 directives, components, and even code conventions. These concepts are still a bit abstract, so we will bring them down to implementation as we move through the rest of the book and work with real code. However, we are on safe footing now to learn about design principles and patterns in the next chapter.

Review questions

To help you consolidate the contents of this chapter, you can use these review questions:

- What is the difference between a library and a framework?
- Why is Vue a "progressive" framework?
- What are single-file components?
- What are some of the most common directives used in Vue development?
- Why are code conventions important?

If you can answer these questions quickly in your mind, you're good to go! If not, you may want to review the chapter briefly to make sure you have the basis to move on.

2
Software Design Principles and Patterns

Software development is fundamentally a *human-intensive discipline*. This means that it requires knowledge of both techniques and technology, but also comprehension of the problem and the ability to make decisions to implement a solution on multiple levels of abstraction. Programming has much to do with how a developer thinks. Over the years, and within each context and language, guidelines and solutions have emerged to solve recurring problems. Knowledge of these *patterns* will help you identify when to apply them and speed your development on a sure footing. On the other hand, *principles* are guiding concepts that should be applied at every stage of the process and have more to do with how you approach the process.

In this chapter, we will take a look at a non-exclusive and non-exhaustive list of principles and patterns that are common in Vue 3 application development.

Principles	Patterns
• Separation of concerns	• Singleton
• Composition over inheritance	• Dependency injection
• Single responsibility	• Observer
• Encapsulation	• Command
• KIC – keep it clean	• Proxy
• DRY – don't repeat yourself	• Decorator
• KISS – keep it simple stupid	• Façade
• Code for the next	• Callbacks
	• Promises

Table 2.1 – The principles and patterns covered in this chapter

Understanding these principles and patterns will help you use the framework more efficiently and more often than not, it will prevent you from "reinventing the wheel". Together with the first chapter, this will conclude the *foundational part* of this book and will give you the basis to follow the practical parts and implementation of application examples in the rest of the book.

What are the principles of software design?

In software development, design principles are high-level conceptual guidelines that should apply to the entire process. Not every project will use the same principles, and these are not mandatory rules to be enforced. They can appear in a project from the architecture down to the **user interface** (**UI**) and the last bit of code. In practice, some of these principles can also influence software attributes, such as maintainability and re-usability.

A non-exclusive list of design principles

Design principles vary somewhat depending on the context, domain, and even the team one may be part of at the time. The principles included in this chapter are, therefore, non-exclusive.

Separation of concerns

This is perhaps the most important principle in software engineering. Separation of concerns implies that a system must be divided into subsystems of elements grouped by their function or service (the **concern**). For example, we can consider the human body as a *system* composed of many subsystems (respiratory, circulatory, digestive, etc.). These, in turn, are integrated by different organs, which are made of tissues, and so forth, down to the smallest cell. Following the same idea in software, an application can be divided into elements grouped by concerns, from the large architecture all the way down to the last function. Without this breakdown of complexity into manageable parts, creating a functional system would be much harder, if not impossible.

In general, the application of this principle starts with the big picture of what the system should be, looks into what it should do to accomplish that, and then breaks it down into manageable working parts.

As an example, here is a crude graphical representation of separation of concerns for a web application. Each box in this diagram identifies a different *concern* that, in turn, can be detailed into smaller functional parts. Even better, you can see how this principle allows you to identify the integrating parts of a system.

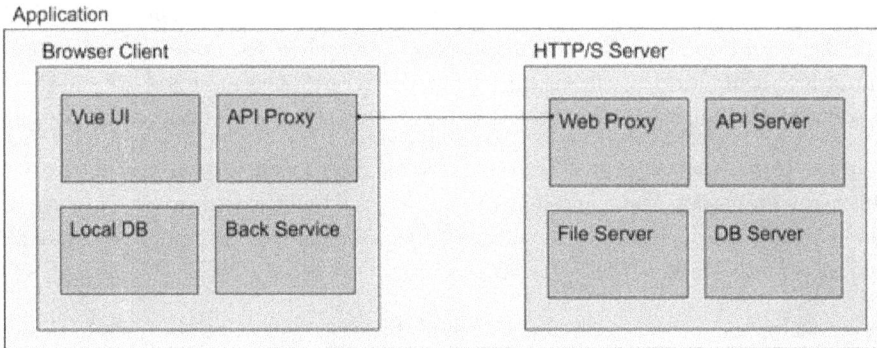

Figure 2.1 – A simple architectural view of a web application showing separation of concerns

If we were to drill down into any of these small boxes within their respective domains, we could still find more concerns to subdivide until we reach an indivisible atomic element (a component or function, for example). This principle has much to do with and benefits from other principles, such as abstraction and single responsibility. We will review them further down the line in this same chapter.

Composition over inheritance

The principle of *composition over inheritance* comes directly from **Object-Oriented Programming (OOP)**. It states that an object should attempt to use other objects' functionality when needed, by referencing or instantiating them instead of creating a large and complex inheritance family tree of classes to add such functionality. Now, JavaScript is fundamentally a *functional* language, even though it supports multiple paradigms, including features from OOP, so this principle applies as well. There is one note of warning for those migrating from OOP into JavaScript, and that is to avoid the temptation to treat JavaScript as a pure OOP language. Doing so could create unnecessary complexity instead of benefiting from the virtues of the language.

In Vue 3, there is no extension or inheritance of components. When we need shared or inherited functionality, we have a nice toolset of options to replace the inheritance paradigm. We will see later how we can comply with this principle by using *composable components* in *Chapter 4, User Interface Composition with Components*.

Single responsibility principle

This principle can be found in OOP as well as in functional programming. Simply put, it states that a class, method, function, or *component* should deal with only one responsibility or functionality. If you have worked in other disciplines and languages, this comes naturally. Multipurpose functions are hard to maintain and tend to grow out of control, especially in a language such as JavaScript, which is loosely typed and highly dynamic. The same concept also applies directly to Vue 3 components. Each component should deal with one specific operation and avoid attempting to do too much by itself. In practice, when a component grows beyond a certain scope, it is best to split it into multiple components or extract the behavior into external modules. There are cases when you may end up with a many-thousand-lines-long component, but in my experience, this is rarely necessary and can and should be avoided. A warning, though, is that too much specificity could also lead to unnecessary complexity.

As an example, let's imagine a sign-in screen that also displays a sign-up option. This approach is common on many sites today. You could include all the functionalities inside just one component, but that would break this principle. A better alternative would be to split the components into at least three components for this task:

- A parent component that handles the UI logic. This component decides when to show/hide the sign-in and sign-up components.
- A child component that handles the sign-in function.
- A child component that handles the sign-up function.

Here is a graphical representation of this configuration:

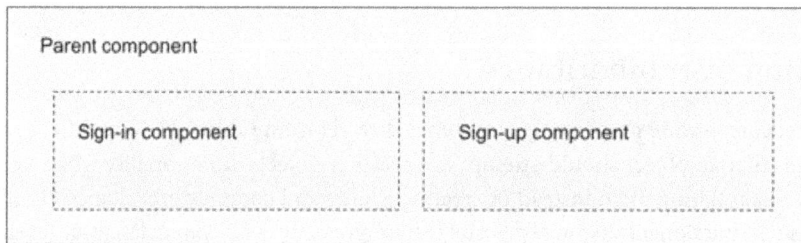

Figure 2.2 – The composition of a sign-in/up interface using multiple components

I think that you can quickly grasp the benefits of this principle. It makes the code easy to manage, maintain, and adapt since web applications have the tendency to mutate and evolve very, very quickly.

> **Best practice tip**
> Give components a single responsibility and functionality. Avoid mammoth monolithic components as much as possible.

Encapsulation

Encapsulation is the notion that you should wrap data and methods to act as a single unit while exposing a well-defined **application programming interface (API)**. Often, this is done in the form of classes, modules, or libraries. JavaScript is not an exception, and it is highly recommended to follow this principle. In Vue 3, this concept applies to not only components but also CSS styles and HTML. The introduction of *single-file components* is a clear example of how the framework promotes this principle in action and how important it is for today's development. With only a few edge-case situations, we should consider the **(UI)** components as black boxes that receive incoming parameters and provide outgoing data. Other components should not be *aware* of their inner workings, only the API. As we build example applications throughout this book, you will see this principle in action.

KIC – keep it clean

This principle refers mainly to the way *you write* code. I should emphasize here that KIC applies directly to two categories that strongly affect web and Vue 3 applications:

- How you format your code
- How you tidy up events and variables

The first item includes the use of code conventions, comments, and indentation to the organization of the code and logical grouping of functions. For example, if you have methods that deal with **create, read, update, and delete (CRUD)** operations, it would be best to place them near each other in the code, rather than spread around the source file. Many **integrated development environments (IDEs)** contain features to collapse or expand the inner code of functions. This helps to quickly review and locate sections in the code with similar logic.

The second part of this principle has to do with memory and reference handling. JavaScript has a very good garbage collector, the function of which is to discard unused data to reclaim memory. However, there are occasions when the algorithm is prevented from freeing up resources because a reference is still pending. If you have worked with other languages, such as C/C++, this issue may sound familiar as you need to manually reserve and release memory when not in use. In JavaScript, if you register a function to listen to an event, it is best to manually deregister it at the appropriate life cycle event of your component when no longer needed. This will prevent memory leaks and waste of memory and also prevent some security risks (which are out of the scope of this book).

We will review the component's life cycle in *Chapter 4, User Interface Composition with Components*, but for now, take the following example as a good application of this principle and keep it as best practice. In this example, we will create a *composable* component to detect when the window size changes, so in the `script setup` section we will find something like this:

1. Registers a function on the window object's resize event during the mounting state.
2. Deregisters the event before the component is unmounted.

Here is the code fragment:

```
<script setup>
    import {onMounted, onBeforeUnmount} from "vue"
    onMounted(()=>{
        window.addEventListener("resize", myFunction)
    })
    onBeforeUnmount(()=>{
        window.removeEventListener("resize", myFunction)
    })
    function myFunction(){
        // Do something with the event here
    }
</script>
```

The `onMounted` and `onBeforeUnmount` functions are part of the Vue 3 framework and are triggered by the appropriate component life cycle event. Here, we attach our function to the `resize` event when the component is mounted to the **Document Object Model** (**DOM**), and we release it just before it is removed. The important concept to remember is to clean up after yourself and *keep it clean*.

DRY – don't repeat yourself

This principle is quite famous, almost to the point of turning into a cliché. Sadly, it is easily forgotten. It is credited to Andrew Hunt and David Thomas, who used it in the book *The Pragmatic Programmer*. It is mostly thought of as *don't write the same thing twice* and is not far off, but it goes beyond that. It encompasses the notion of avoiding redundancy in the process as well as in the logic of the application. The core idea is that each process that executes business logic should exist in only one place in your entire application.

For example, most web applications have some asynchronous connection with a server through the use of an API. There may also be multiple elements in the application that will use or need to use this remote computer/server communication. If you were going to code the entire code/logic to communicate with the server in each component, we would end up with not only duplication of code but also application logic. Maintaining such a system would open up the door to an amazing number of negative side effects and security concerns, poor user experience, and much more. According to this principle, a better approach is to abstract all communication code related to the server API into a single module, or class. In practice, in JavaScript this can even be delegated to a web worker in a separate thread. We will explore this implementation later in *Chapter 8, Multithreading with Web Workers*.

As a rule of thumb, if you see yourself writing kind-of-the-same-code" in different components or classes, it is a clear opportunity to abstract the functionality into its own module or component.

KISS – keep it simple and short

This principle is not exclusive to the software design realm. It was coined by the US Navy back in the '60s (according to Wikipedia, `https://en.wikipedia.org/wiki/KISS_principle`). The idea is pure common sense: it is better to build simple, small functional parts that work together than attempt to create a big and complex program in one go. Also, algorithms should be implemented in the most simple and efficient way. In web development, this principle is essential. Modern web applications are composed of hundreds of working parts spread over multiple computers, servers, and environments. The more complex a system or code implementation is, the harder it is also to maintain and adapt.

There is a warning, though. Keeping things simple does not mean over-simplification or unnecessary segregation. Too many small parts can introduce unnecessary complexity in the system. Applying the KISS principle means staying in that sweet middle point where things are manageable and easy to understand.

Code for the next

This principle is the idea that you should make your code readable and easy to understand for someone else besides you. Naming conventions, logic flow, and inter-line comments are all part of this. Not only for the case when you may need to delegate your code to another but also when you come back in a year or two to the same code. The last thing you want to do is to waste time thinking about what the past inexperienced you did with that clever line of spaghetti code Smart developers code as if they were going to teach somebody else, simply and elegantly. Especially if you are using or contributing to open-source code, this principle is vital for group collaboration. In this case, it is worth mentioning the *Boy Scout Principle,* which is similar but applies in groups. It states that when you find a hard-to-read or "spaghetti" code, you refactor it to make it clean.

> **Best practice tip**
>
> Keep your code clean with on-source comments and documentation explaining your logic, as if teaching somebody else. More often than not, you will be teaching yourself.

Design principles apply to many different scenarios, some beyond the practice of software development. It is important to consider them until they become second nature. In general, the application of these and other principles, together with the application of design patterns, make an important mark on your professional development.

What is a software design pattern?

In software development, it is common for certain processes and tasks to appear in multiple projects, in one way or another, or with some degree of variation. A *design pattern* is a proven solution for such similar problems. It does not prescribe code but acts like a reasoning template, an approach that has been abstracted independent of the implementation to be reused and adapted to particular circumstances. In practice, there is plenty of room for creativity to apply a pattern. Entire books have been dedicated to this subject and provide more detail than the scope of this book allows. In the following pages, we will take a look at what I consider to be the most recurrent patterns to keep in mind for Vue 3 applications. Even though we see them in isolation for the purposes of studying them, the reality is that often the implementation overlaps, mixes, and encapsulates multiple patterns in a single piece of code. For example, you can use a **singleton** to act as a **decorator** and a **proxy** to simplify or alter the communication between services in your application (we will do this quite often, actually, and the full code can be seen in *Chapter 8, Multithreading with Web Workers*).

Design patterns can also be understood as software engineering and development *best practices*. And the opposite of that, *bad practice*, is often referred to as an **anti-pattern**. Anti-patterns are "solutions" that, even though they fix an issue in the short term, create problems and bad consequences along the line. They generate the need to work around the problem and destabilize the whole structure and implementation.

Let's now view a list of patterns that should be part of your toolbox for Vue 3 projects.

A quick reference list of patterns

Patterns are classified according to the type of function or problem they solve. There are plenty of patterns according to the context, language, and architecture of a system. Here is a non-exclusive list of patterns that we will use throughout this book and that, in my experience, are more likely to appear in Vue applications:

- **Creational patterns**: These deal with the approach to creating classes, objects, and data structures:

    - Singleton pattern

    - Dependency injection pattern

    - Factory pattern

- **Behavioral patterns**: These deal with communication between objects, components, and other elements of the application:

    - Observer pattern

    - Command pattern

- **Structural patterns**: These provide templates that affect the design of your application and the relationship between components:

  - Proxy pattern

  - Decorator pattern

  - Façade pattern

- **Asynchronous patterns**: These deal with data and process flow with asynchronous requests and events in single-threaded applications (heavily used in web applications):

  - Callbacks pattern

  - Promises pattern

Not by any means this list of patterns is exclusive. There are many more patterns and classifications, and a full library is dedicated to this subject. It is worth mentioning that the description and application for some of these may differ from one literature to another and there is some overlapping depending on the context and implementation.

With that introduction to design patterns, let's look at them in detail with examples.

The singleton pattern

This is a very common pattern in JavaScript and perhaps one of, if not the most important. The basic concept defines that one object's instance must only exist once in the entire application, and all references and function calls are done through this object. A singleton can act as a gateway to resources, libraries, and data.

When to use it

Here is a short rule of thumb to know when to apply this pattern:

- When you need to make sure a resource is accessed through only one gateway, for example, the global application state

- When you need to encapsulate or simplify behavior or communications (used in conjunction with other patterns). For example, the API access object.

- When the *cost* of multiple instantiations is detrimental. For example, the creation of web workers.

Implementations

There are many ways that you can apply this pattern in JavaScript. In some cases, the implementation from other languages is migrated to JavaScript, often following Java examples with the use of a `getInstance()` method to obtain the singleton. However, there are better ways to implement this pattern in JavaScript. Let's see them next.

Method 1

The simplest way is through a module that exports a plain object literal or a **JavaScript Object Notation (JSON)**, which is a static object:

./chapter 2/singleton-json.js

```
const my_singleton={
    // Implementation code here...
}
export default my_singleton;
```

You then can import this module into other modules and still always have the same object. This works because bundlers and browsers are smart enough to avoid the repetition of imports, so once this object has been brought in the first time, it will ignore the next requests. When not using a bundler, the ES6 implementation of JavaScript also defines that modules are singletons.

Method 2

This method creates a class and then, on the first instantiation, saves the reference for future calls. In order for this to work, we use a variable (traditionally called `_instance`) from the class and save the reference to the instance in the constructor. In the following calls, we check whether the `_instance` value exists, and if so, return it. Here is the code:

./chapter 2/singleton-class.js

```
class myClass{
    constructor(){
        if(myClass._instance){
            return myClass._instance;
        }else{
            myClass._instance=this;
        }
        return this;
    }
}
export default new myClass()
```

This second method may be more familiar to other language developers. Notice how we are also exporting a new instance of the class and not the class directly. This way, the invoker will not have to remember to instantiate the class every time, and the code will be the same as in *method 1*. This use case is something that needs to be coordinated with your team to avoid different implementations.

The invoker then can call methods from each one directly (assuming the singleton has a function/method called myFunction()):

./chapter 2/singleton-invoker.js

```
import my_method1_singleton from "./singleton-json";
import my_method2_singleton from "./singleton-class";
console.log("Look mom, no instantiation in both cases!")
my_method1_singleton.myFunction()
my_method2_singleton.myFunction()
```

The singleton pattern is extremely useful, though it rarely exists *in isolation*. Often, we use singletons to wrap the implementation of other patterns and make sure we have a single point of access. In our examples, we will use this pattern quite often.

The dependency injection pattern

This pattern simply states that the dependencies for a class or function are provided as inputs, for example, as parameters, properties, or other types of implementations. This simple statement opens a very wide range of possibilities. Let's take, for example, a class that works with the browser's **IndexedDB API** through an abstraction class. We will learn more about the IndexedDB API in *Chapter 7, Data Flow Management*, but for now, just concentrate on the dependency part. Consider that the dbManager.js file exposes an object that handles the operations with the database, and the projects object deals with CRUD operations for the projects table (or collection). Without using dependency injection, you will have something like this:

./chapter 2/dependency-injection-1.js

```
import dbManager from "dbManager"
const projects={
    getAllProjects(){
        return dbManager.getAll("projects")
    }
}
export default projects;
```

The preceding code shows a "normal" approach, where we import the dependencies at the beginning of the file and then use them in our code. Now, let's tweak this same code to use dependency injection:

./chapter 2/dependency-injection-2.js

```
const projects={
    getAllProjects(dbManager){
        return dbManager.getAll("projects")
    }
}
export default projects;
```

As you can see, the main difference is that dbManager is now passed as a parameter to the function. This is what is called **injection**. This opens up many ways to manage dependencies and, at the same time, pushes the hardcoding of dependencies up the implementation tree. This makes this class highly reusable, at least for as long as the dependency respects the expected API.

The preceding example is not the only way to inject a dependency. We could, for example, assign it to a property for the object's internal use. For example, if the projects.js file was implemented using the property approach instead, it would look like this:

./chapter 2/dependency-injection-3.js

```
const projects={
    dbManager,
    getAllProjects(){
        return this.dbManager.getAll("projects")
    }
}
export default projects;
```

In this case, the invoker of the object (a singleton, by the way) needs to be aware of the property and assign it before calling on any of its functions. Here is an example of how that would look:

./chapter 2/dependency-injection-4.js

```
import projects from "projects.js"
import dbManager from "dbManager.js"
projects.dbManager=dbManager;
projects.getAllProjects();
```

But this approach is not recommended. You can clearly see that it breaks the principle of encapsulation, as we are directly assigning a property for the object. It also doesn't feel like clean code even though it is valid code.

Passing the dependencies one function at a time is also not recommended. So, what is a better approach? It depends on the implementation:

- In a class, it is convenient to require the dependencies in the constructor (and if not found, throw an error)

- In a plain JSON object, it is convenient to provide a function to set the dependency explicitly and let the object decide how to use it internally

This last approach is also recommended for passing a dependency after the instantiation of an object when the dependency is not ready at the time of implementation

Here is a code example for the first point mentioned in the preceding list:

./chapter 2/dependency-injection-5.js

```
class Projects {
    constructor(dbManager=null){
        if(!dbManager){
            throw "Dependency missing"
        }else{
            this.dbManager=dbManager;
        }
    }
}
```

In the constructor, we declare the expected parameter with a default value. If the dependency is not provided, we throw an error. Otherwise, we assign it to an internal private attribute for the use of the instance. In this case, the invoker should look like this:

```
// Projects are a class
import Projects from "projects.js"
import dbManager from "dbManager.js"
try{
    const projects=new Projects(dbManager);
}catch{
    // Error handler here
}
```

In an alternative implementation, we could have a function that basically does the same by receiving the dependency and assigning it to a private attribute:

```
import projects from "projects.js"
import dbManager from "dbManager.js"
projects.setDBManager(dbManager);
```

This approach is better than directly assigning the internal attribute, but you still need to remember to do the assignment before using any of the methods in the object.

> **Best practice note**
> Whatever approach you use for dependency injection, remain constant throughout your code base.

You may have noticed that we have mainly been focusing on objects. As you may have already guessed, passing a dependency to a function is just the same as passing another parameter, so it does not deserve special attention.

This example has just moved the dependency implementation responsibility up to another class in the hierarchy. But what if we implement a singleton pattern to handle all or most of the dependencies in our application? This way, we could just delegate the loading of the dependencies to one class or object at a determined point in our application life cycle. But how do we implement such a thing? We will need the following:

- A method to register the dependency
- A method to retrieve the dependency by name
- A structure to keep the reference to each dependency

Let's put that into action and create a very *naive* implementation of such a singleton. Please keep in mind that this is an academic exercise, so we are not considering error checking, de-registration, or other considerations:

./chapter 2/dependency-injection-6.js

```
const dependencyService={                          //1
    dependencies:{},                               //2
    provide(name, dependency){                     //3
        this.dependencies[name]=dependency         //4
        return this;                               //5
    },
    inject(name){                                  //6
        return this.dependencies[name]??null;      //7
    }
}
export default dependencyService;
```

With this bare minimum implementation, let's look at each line by the line comment:

1. We create a simple JavaScript object literal as a singleton.
2. We declare an empty object to use as a dictionary to hold our dependencies by name.

3. The `provide` function lets us register a dependency by name.

4. Here, we just use the name as the field name and assign the dependency passed by argument (notice we are not checking pre-existing names, etc.).

5. Here, we return the source object, mainly for convenience so we can chain the invocation.

6. The `inject` function will take the name as registered in the `provide` function.

7. We return the dependency or `null` if not found.

With that singleton on board, we can now use it across our application to distribute the dependencies as needed. For that, we need a parent object to import them and populate the service. Here is an example of how that might look:

./chapter 2/dependency-injection-7.js

```
import dependencyService from "./dependency-injection-6"
import myDependency1 from "myFile1"
import myDependency2 from "myFile2"
import dbManager from "dbManager"
dependencyService
    .provide("dependency1", myDependency1)
    .provide("dependency2", myDependency2)
    .provide("dbManager", dbManager)
```

As you can see, this module has hard-coded dependencies, and its work is to load them into the `dependencyService` object. Then, the dependent function or object needs only to import the service and retrieve the dependency it needs by the registration name like this:

```
import dependencyService from "./dependency-injection-6"
const dbManager=dependencyService.inject("dbManager")
```

This approach does create a tight coupling between components but is here as a reference. It has the advantage that we can control all the dependencies in a single location so that the maintenance benefits could be significant. The choice of names for the methods of the `dependencyService` object was not random either: these are the same used by Vue 3 inside the component's hierarchy. This is very useful for implementing some User Interface design patterns. We will see this in more detail in *Chapter 4, User Interface Composition with Components* and *Chapter 7, Data Flow Management*.

As you can see, this pattern is very important and is implemented in Vue 3 with the `provide`/`inject` functions. It's a great addition to our toolset, but there is more still. Let's move on to the next one.

The factory pattern

The factory pattern provides us with a way to create objects without creating a direct dependency. It works through a function that, based on the input, will return an instantiated object. The use of such an implementation will be made through a common or standard interface. For example, consider two classes: `Circle` and `Square`. Both implement the same `draw()` method, which draws the figure to a canvas. Then, a `factory` function would work something like this:

```
function createShape(type){
    switch(type){
        case "circle": return new Circle();
        case "square": return new Square();
}}
let
    shape1=createShape("circle"),
    shape2=createShape("square");
shape1.draw();
shape2.draw();
```

This method is quite popular, especially in conjunction with other patterns, as we will see multiple times in this book.

The observer pattern

The observer pattern is very useful and one of the basis of a reactive framework. It defines a relationship between objects where one is being observed (the **subject**) for changes or events, and other(s) are notified of such changes (the **observers**). The observers are also called **listeners**. Here is a graphical representation:

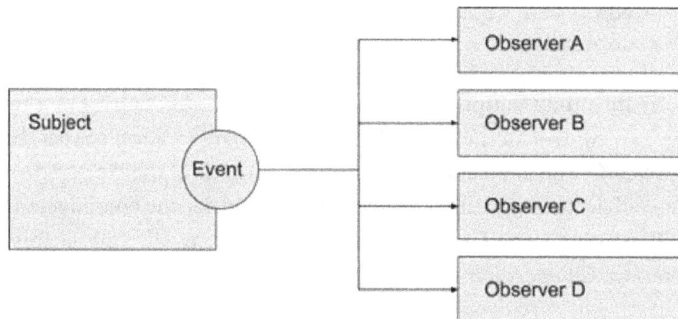

Figure 2.3 – The subject emits an event and notifies the observers

As you can see, the subject emits an event to notify the observers. It is for the subject to define what events and parameters it will **publish**. Meanwhile, the observers **subscribe** to each event by registering a function with the publisher. This implementation is why this pattern is often referred to as the **pub/sub** pattern, and it can have several variations.

When looking into the implementation of this pattern, it is important to notice the cardinality of the publication: 1 event to 0..N observers (functions). This means that the subject must implement, on top of its main purpose, the functionality to publish events and keep track of the subscribers. Since this would break a principle or two in the design (separation of concerns, single responsibility, etc.), it is common to extract this functionality into a middle object. The previous design then changes to add a middle layer:

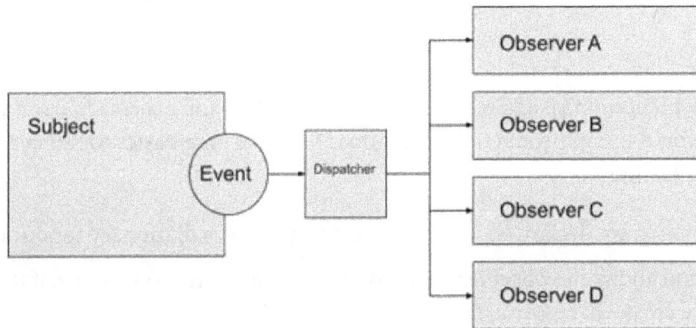

Figure 2.4 – An observer implementation with a dispatcher middle object

This middle object, sometimes referred to as an "**event dispatcher** encapsulates the basic functionality to register observers, receive events from the subject, and dispatch them to the observers. It also does some clean-up activities when an observer is no longer observing Let's put these concepts into a simple and naive implementation of an event dispatcher in plain JavaScript:

./chapter 2/Observer-1.js

```
class ObserverPattern{
constructor(){
    this.events={}                                    //1
}
on(event_name, fn=()=>{}){                            //2
    if(!this.events[event_name]){
        this.events[event_name] = []
    }
    this.events[event_name].push(fn)                  //3
}
emit(event_name, data){                               //4
```

```
    if(!this.events[event_name]){
        return
    }
for(let i=0, l=this.events[event_name].length; i<l; i++){
    this.events[event_name][i](data)
}
}
off(event_name, fn){                                          //5
    let i=this.events[event_name].indexOf(fn);
    if(i>-1){
        this.events[event_name].splice(i, 1);
    }
}
}
```

The preceding implementation is, again, naive. It doesn't contain the necessary error and edge case handling that you would use in production, but it does have the bare basics for an event dispatcher. Let's look into it line by line:

1. In the constructor, we declare an object to use internally as a dictionary for our events.

2. The on method allows the observers to register their functions. In this line, if the event is not initialized, we create an empty array.

3. In this line, we just push the function to the array (as I said, this is a naive implementation, as we don't check for duplicates, for example).

4. The emit method allows the subject to publish an event by its name and pass some data to it. Here, we run over the array and execute each function passing the data we received as a parameter.

5. The off method is necessary to deregister the function once it is not used (see the *keep it clean* principle, earlier in this chapter).

In order for this implementation to work, every observer and the subject need to reference the same implementation of the ObserverClass. The easiest way to secure this is to implement it through a *singleton pattern*. Once imported, each observer registers with the dispatcher with this line:

```
import dispatcher from "ObserverClass.js"      //a singleton
dispatcher.on("event_name", myFunction)
```

Then, the subject emits the event and passes the data with the following lines:

```
import dispatcher from "ObserverClass.js"      //a singleton
dispatcher.emit("event_name", data)
```

Finally, when the observer no longer needs to watch the subject, it needs to clean up the reference with the `off` method:

```
dispatcher.off("event_name", myFunction)
```

There are a good number of edge cases and controls that we have not covered here, and rather than reinventing the wheel, I suggest using a ready-made solution for these cases. In our book, we will use one named `mitt` (`https://www.npmjs.com/package/mitt`). That has the same methods as in our example. We will see how to install packaged dependencies in *Chapter 3, Setting up a Working Project*.

The command pattern

This pattern is very useful and easy to understand and implement. Instead of executing a function right away, the basic concept is to create an object or structure with the information necessary for the execution. This data package (the **command**) is then delegated to another object that will perform the execution according to some logic to handle it. For example, the commands can be serialized and queued, scheduled, reversed, grouped together, and transformed. Here is a graphical representation of this pattern with the necessary parts:

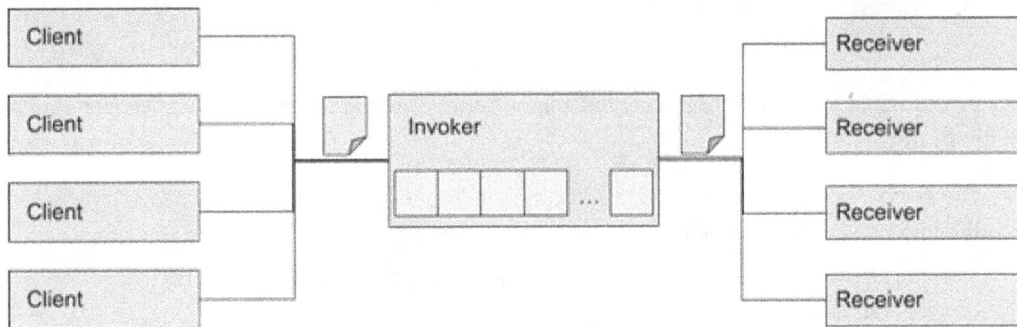

Figure 2.5 – A graphical implementation of the command pattern

The diagram shows how the clients submit their commands to the **Invoker**. The invoker usually implements some sort of queue or task array to handle the commands and then routes the execution to the proper **Receiver**. If there is any data to return, it also returns it to the proper client. It is also common that the invoker attaches additional data to the command to keep track of clients and receives, especially in the case of asynchronous executions. It also provides a single point of "entry" to the receivers and decouples the "clients" from them.

Let's again work on a naive implementation of an `Invoker` class:

./chapter 2/Command-1.js

```js
class CommandInvoker{
    addCommand(command_data){                          //1
        // .. queue implementation here
    }
    runCommand(command_data){                          //2
        switch(command_data.action){                   //3
            case "eat":
                // .. invoke the receiver here
                break;
            case "code":
                // .. invoke the receiver here
                break;
            case "repeat":
                // .. invoke the receiver here
                break;
        }
    }
}
```

In the preceding code, we have implemented a bare-bones example of what an `Invoker` should have line by line:

1. The `Invoker` exposes a method to add commands to the object. This is only necessary when the commands will be somehow queued, serialized, or processed according to some logic.

2. This line executes the command according to the `action` field contained in the command_ data parameter.

3. Based on the `action` field, the *invoker* routes the execution to the proper receiver.

There are many ways to implement the logic for routing the execution. It is important to notice that this pattern can be implemented on a larger scale depending on the context. For example, the invoker might not even be in the web client application and be on the server or on a different machine. We will see an implementation of this pattern in *Chapter 8, Multithreading with Web Workers*, where we use this pattern to process tasks between different threads and unload the main thread (where Vue 3 runs).

The proxy pattern

The definition for this pattern comes directly from its name, as the word "proxy" means something or someone who acts on behalf of another as if it was the same. That is a mouthful, but it will make you remember it. Let's look into an example to clarify how this works. We will need at least three entities (components, objects, etc.):

- A **client** entity that needs to access the API of a target entity
- A **target** entity that exposes a well-known API
- A **proxy** object that sits in between and exposes the same API as the target while at the same time intercepting every communication from the client and relaying it to the target

We can graphically represent the relationship between these entities in this way:

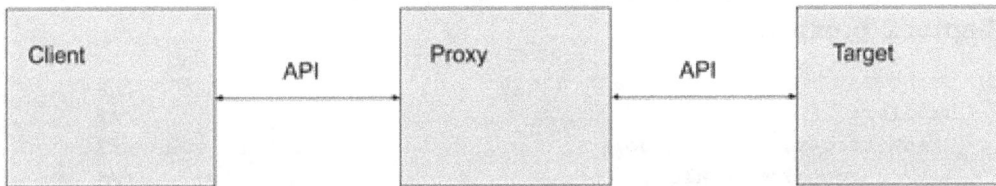

Figure 2.6 – The proxy object exposes the same API as the target

The key factor for this pattern is that the proxy behaves and exposes the same API as the target, in such a way that the client does not know or doesn't need to know that it is dealing with a proxy and not the target object directly. So, why would we want to do such a thing? There are many good reasons, such as the following:

- You need to maintain the original unmodified API, but at the same time:

  - Need to process the inputs or outputs for the client

  - Need to intercept each API call to add internal functionality, such as maintenance operations, performance improvements, error checking, and validation

  - The target is an expensive resource, so a proxy could implement logic to leverage their operations (for example, a cache)

- You need to change the client or the target but can't modify the API
- You need to maintain backward compatibility

There are more reasons that you may come across, but I hope that by now you can see how this can be useful. Being a pattern, this template can be implemented on multiple levels, from a simple object proxy to a full application or server. It is quite common when performing partial upgrades of a system or application. On a lower level, JavaScript even natively includes a constructor for proxying objects that Vue 3 uses internally to create reactivity.

In *Chapter 1, The Vue 3 Framework*, we reviewed the options for reactivity with the ref() but this new version of Vue also includes another alternative for complex structures, called reactive(). The first one uses pub/sub methods (the observer pattern!), but the latter uses native proxy handlers (this pattern!). Let's look into an example of how this native implementation may work with a naive partial implementation.

In this simple example, we will make an object with reactive properties automatically convert Celsius degrees to and back from Fahrenheit using a Proxy object:

./chapter 2/proxy-1.js

```
let temperature={celsius:0,fahrenheit: 32},          //1
    handler={                                        //2
      set(target, key, value){                       //3
          target[key]=value;                         //4
      switch(key){
        case "celsius":
            target.fahrenheit=calculateFahrenheit(value);   //5
            break;
        case "fahrenheit":
            target.celsius=calculateCelsius(value);
          }
      },
      get(target, key){
          return target[key];                        //6
      }
    },
    degrees=new Proxy(temperature, handler)          //7
// Auxiliar functions
function calculateCelsius(fahrenheit){
    return (fahrenheit - 32) / 1.8
}
function calculateFahrenheit(celsius){
    return (celsius * 1.8) + 32
}
degrees.celsius=25                                   //8
console.log(degrees)
// Prints in the console:
// {celsius:25, fahrenheit:77}                       //9
```

Let's review the code line by line to see how this works:

1. In this line, we declare the `temperature` object, which is going to be our target to be proxied. We initialize its two properties with an equal converted value.

2. We declare a `handler` object, which will be our proxy for the temperature object.

3. The `set` function in the proxy handler receives three arguments: the target object, the key referred to, and the value attempted to be assigned. Notice that I say "attempted", as the operation has been intercepted by the proxy.

4. On this line, we perform the assignment as intended to the object property. Here, we could have done other transformations or logic, such as validation or raised an event (the observer pattern again!).

5. Notice how we use a switch to filter the property names that we are interested in. When the key is `celsius`, we calculate and assign the value in Fahrenheit. The opposite happens when we receive an assignment for `fahrenheit` degrees. This is where the **reactivity** comes into play.

6. For the `get` function, at least in this example, we just specifically return the value requested. In the way this is implemented, it would be the same as if we skip the `getter` function. However, it is here as an example that we could operate and transform the value to be returned as this operation is also intercepted.

7. Finally, in line 7, we declare the `degrees` object as the proxy for `temperature` with the handler.

8. On this line, we test the reactivity by assigning a value in *Celsius* to the member of the `degrees` object, just like we normally would to any other object.

9. When we print the `degrees` object to the console, we notice that the `fahrenheit` property has been automatically updated.

This is a rather limited and simple example of how the native `Proxy()` constructor works and applies the pattern. Vue 3 has a more complex approach to reactivity and tracking dependencies, using the proxy and observer patterns. However, this gives us a good idea of what approach is happening behind the scenes when we see the HTML updated live in front of our very eyes.

The concept of proxying between a client and a target is also related to the next two patterns: the *decorator and the façade patterns* since they are also a sort of proxy implementation. The distinguishing key factor is that the proxy retains the same API as the original target object.

The decorator pattern

This pattern may, at first sight, seem very similar to the *proxy pattern*, and indeed it is, but it adds a few distinctive features that set it apart. It does have the same moving parts as the proxy, meaning there is a **Client**, a **Target**, and a **Decorator** in between that implements the same interface as the target (yes, just like in the proxy pattern). However, while in the *Proxy pattern* the intercepted API calls mainly deal with the data and internal maintenance ("housekeeping"), the decorator augments the functionality of the original object to do more. This is the defining factor that separates them.

In the proxy example, notice how the additional functionality was an *internal reactivity* to keep the degrees in each scale synchronized. When you change one, it internally and automatically updates the other. In a decorator pattern, the proxy object performs additional operations before, during, or after executing the API call to the target object. Just like in the proxy pattern, all of this is transparent for the client object.

For example, building on the previous code, imagine that now we want to log each call to the API of a certain target while keeping the same functionality. Graphically, it would look like this:

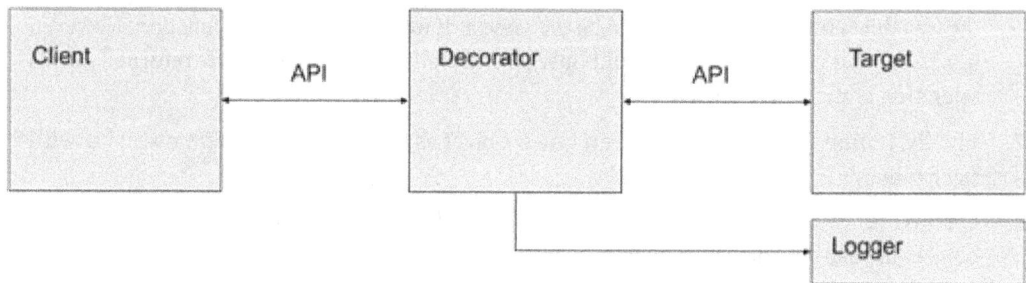

Figure 2.7 – An example of a decorator that augments the target with a logging feature

Here, what was first a simple proxy, now by the mere act of performing a humble logging call, has now become a decorator. In the code, we only need to add this line before the end of the set() method (assuming there is also a function named getTimeStamp()):

```
console.log(getTimeStamp());
```

Of course, this is a simple example just to make a point. In the real world, decorators are very useful for adding functionality to your application without having to rewrite the logic or significant portions of your code. On top of this, decorators can be *stackable* or *chainable*, meaning that you can create "decorators for decorators" if needed, so each one will represent one step of added functionality that would maintain the same API of the target object. And just like that, we are beginning to step into the boundaries of a **middleware pattern**, but we will not cover it in this book. Anyway, the idea behind that other pattern is to create layers of middleware functions with a specified API, each one that performs one action, but with the difference that any step can decide to abort the operation, so the target may or may not be called. But that is another story... let's get back to decorators.

Previously in this book, we mentioned that Vue 3 components do not have inheritance like plain JavaScript classes implemented by extending from one another. Instead, we can use the decorator pattern on components to add functionality or change the visual appearance. Let's look at a brief example now, as we will see components and UI design in detail in *Chapter 4, User Interface Composition with Components*.

Consider that we have the simplest of components that displays a humble h1 tag with a title that receives the following as input:

./chapter 2/decorator-1.vue

```
<script setup>
    const $props=defineProps(['label'])          //1
</script>
<template>
    <h1>{{$props.label}}</h1>                      //2
</template>
<style scoped></style>
```

In this simple component, we declare a single input named label in line //1. Don't worry about the syntax for now, as we will see this in detail in *Chapter 4, User Interface Composition with Components*. On line //2, we are interpolating the value plainly inside the h1 tags just as expected.

So, to create a decorator for this component we need to apply the following simple rules:

- It has to act on behalf of the component (object)

- It has to respect the same API (inputs, outputs, function calls, etc.)

- It has to augment the functionality or visual representation before, after, or during the execution of the target API

With that in mind, we can create a decorator component that intercepts the label attribute, changes it a bit, and also modifies the visual appearance of the target component:

./chapter 2/decorator-2.vue

```
<script setup>
    import HeaderH1 from "./decorator-1.vue"
    const $props=defineProps(['label'])          //1
</script>
<template>
    <div style="color: purple !important;">        //2
        <HeaderH1 :title="$props.label+'!!!'">     //3
        </HeaderH1>
    </div>
</template>
```

In this code, in line //1, you can see that we keep the same interface as the target component (that we imported in the previous line), and then in line //2, we modify (augment) the color attribute and in line //3 we are also modifying the data passed to the target component by adding three exclamation marks. With those simple tasks, we have kept the conditions to build a decorator pattern extrapolated to Vue 3 components. Not bad at all.

Decorators are very useful, but there is still one more proxy-like pattern that is also very common and handy: the façade pattern.

The façade pattern

By now, you may have seen the progressive pattern in these, well, patterns. We started with a proxy to act on behalf of another object or entity, we augmented it with the use of decorators while keeping the same API, and now is the turn for the façade pattern. Its job is, in addition to the functions of a proxy and decorator, to simplify the API and hide the large complexity behind it. So, a façade sits between a client and a target, but now the target is highly complex, being an object or even a system or multiple subsystems. This pattern is also used to change the API of an object or to limit the exposure to the client. We can picture the interactions as follows:

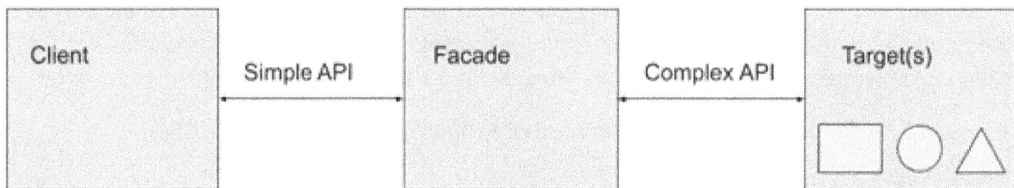

Figure 2.8 – A façade object simplifying the interaction with a complex API or system

As you can see, the main purpose of the façade is to offer a simpler approach to a complex interaction or API. We will use this pattern many times during our examples to simplify native implementations in the browser with more developer-friendly approaches. We will use libraries to encapsulate the use of IndexedDB and create our own simplified communication with web workers in *Chapter 8, Multithreading with Web Workers*.

Needless to say, you will have seen this pattern in action before, as it is one of the foundational concepts of modern technology. *Hiding complexity* behind a simple interface (API) is all around us and is a big part of web development. After all, the entire internet is extremely complicated, with thousands of moving parts, and the technology that makes up web pages is close to magic. Without this pattern, we would still be programming with zeros and ones.

In practice, you will add layers of simplification to your own applications to break down complexity. One way to do it is to use third-party libraries that provide a simplified interface. In the following chapters, we will use some of these, such as the following:

- **Axios**: To handle all **Asynchronous JavaScript and XML (AJAX)** communications with the server

- **DexieDB**: To handle the API to IndexedDB (the browser's local database)

- **Mitt**: To create event pipelines (we mentioned this in the Observer pattern)

- **Vue 3**: To create amazing UIs

In general, there are façade libraries for most of the native implementations of web technologies, which are well battle tested. Developers are very good at simplifying these and sharing the code with others, thanks to the open source movement. Still, when using other people's modules, make sure they are "safe." Don't reinvent the wheel, and don't repeat yourself But now, it is time to move on to the next pattern in our list.

The callback pattern

The callback pattern is easy to understand. It applies when an operation needs to be executed after a **synchronous** or **asynchronous** operation has finished. For this, the function invocation includes, as one of the parameters, a function to be executed when the operations are completed. Having said that, we need to distinguish between the following two types of code flow:

- Synchronous operations are executed one after another in sequential order. It is the basic code flow, top to bottom.

- Asynchronous operations are executed out of the normal flow once invoked. Their length is uncertain, as well as their success or failure.

It is for asynchronous cases that the *callback pattern* is especially useful. Think, for example, of a network call. Once invoked, we don't know how long it will take to get an answer from the server and whether it will succeed, fail, or throw an error. If we didn't have asynchronous operations, our application would be *frozen*, waiting until a resolution happens. That would not be a good user experience, even though it would be computationally correct.

One important feature in JavaScript is that, being single-threaded, asynchronous functions don't block the main thread allowing the execution to continue. This is important since the rendering functions of the browser run on the same thread. However, this is not free as they do consume resources, but they won't freeze the UI, at least in theory. In practice, it will depend on a number of factors heavily influenced by the browser environment and the hardware. Still, let's stick to the theory.

Let's see an example of a synchronous callback function and turn it asynchronous. The example function is very simple: we will calculate the Fibonacci value of a given number using the callback pattern. But first, a refresher on the formula for the calculation:

```
F(0)=0
F(1)=1
F(n)=F(n-1)+F(n-2), with n>=2
```

So, here is a JavaScript function that applies the formula and receives a callback to return the value. Notice that this function is synchronous:

./chapter 2/callback-1.js - Synchronous Fibonacci

```
function FibonacciSync(n, callback){
    if(n<2){
        callback(n)
    } else{
        let pre_1=0,pre_2=1,value;
        for(let i=1; i<n; i++){
            value=pre_1+pre_2;
            pre_1=pre_2;
            pre_2=value;
        }
        callback(value)
    }
}
```

Notice how instead of returning the value with `return`, we are passing it as a parameter to the `callback` function. When is it useful to use such a thing? Consider these simple examples:

```
FibonacciSync(8, console.log);
// Will print 21 to the console
FibonacciSync(8, alert)
// Will show a modal with the number 21
```

Just by replacing the callback function, we can considerably alter how the result is presented. However, the example function has a fundamental flaw affecting the user experience. Being synchronous, the calculation time is proportional to the parameter passed: the larger n, the more time it will take. With a sufficiently large number, we can easily hang up the browser, but also, much before that, we can freeze the interface. You can test that the execution is synchronous with the following snippet:

```
console.log("Before")
FibonacciSync(9, console.log)
console.log("After")
// Will output
```

```
// Before
// 34
// After
```

To turn this simple function into an asynchronous function, you can simply wrap the logic inside a `setImmediate` call. This will take the execution out of the normal workflow. The new function now looks like this:

```
function FibonacciAsync(n, callback){
    setImmediate(()=>{
        if (n<2){
            callback(n)
        } else{
            let pre_1=0,pre_2=1,value;
            for(let i=1; i<n; i++){
                value=pre_1+pre_2;
                pre_1=pre_2;
                pre_2=value;
            }
            callback(value);
        }
    })
}
```

As you can see, we use an arrow function to wrap up the code without any modifications. Now, see the difference when we execute the same snippet as before with this function:

```
console.log("Before")
FibonacciAsync(9, console.log)
console.log("After")
// Will output
// Before
// After
// 34
```

As you can see by the output, the snippet outputs `After` before `34`. This is because our asynchronous operation has been taken out of the normal flow as expected. When calling an asynchronous function, the execution *does not wait* for a result and continues executing the next instruction. This can be confusing at times but is very powerful and useful. However, the pattern does not prescribe how to handle errors or failed operations or how to chain or sequentially run multiple calls. There are different ways to deal with those cases, but they are not part of the pattern. There is another way to handle asynchronous operations that offers more flexibility and control: *promises*. We will see this next, and in most cases, you can use either pattern interchangeably. I say, "in most cases," not all!

The promise pattern

The *promises pattern* is made primarily to deal with asynchronous operations. Just like with callbacks, the invocation of a promised function takes the execution out of the normal flow, but it returns a special object called Promise. This object exposes a simple API with three methods: then, catch, and finally:

- The then method receives two callback functions, traditionally called resolve and reject. They are used in the asynchronous code to return a successful value (resolve) or a failed or negative value (reject).

- The catch method receives an error parameter and is triggered when the process throws an error and the execution is interrupted.

- The finally method executes in either case and receives a callback function.

While a promise is running, it is said to be in an *indeterminate* state until it is resolved or rejected. There is no time limit for how long a promise will wait in this state, something that makes it especially useful for lengthy operations such as network calls and **inter-process communication (IPC)**.

Let's see how to implement the previous example with the Fibonacci series using promises:

```
function FibonacciPromise(n) {
    return new Promise((resolve, reject) => {          //1
        if (n < 0) {
            reject()                                    //2
        } else {
            if (n < 2) {
                resolve(n)                              //3
            } else {
                let pre_1 = 1, pre_2 = 1, value;
                for (let i = 2; i < n; i++) {
                    value = pre_1 + pre_2;
                    pre_1 = pre_2;
                    pre_2 = value;
                }
                resolve(value);
            }
        }
    })
}
```

At first sight, it is easy to see that the implementation has changed a bit. We start on line `//1` by immediately returning a `new Promise()` object. This constructor receives a callback function, that will, in turn, receive two callbacks named `resolve()` and `reject()`. We need to use these in our logic to return a value in case of success (`resolve`) or failure (`reject`). Also notice that we don't have to wrap our code in a `setImmediate` function, as a promise is by nature asynchronous. We now check for negative numbers and then reject the operation in that case (line `//2`). The other change we make is to replace the `callback()` invocation for `resolve()` in lines `//3` and `//4`.

The invocation now also changes:

```
console.log("Before")
FibonacciPromise(9).then(
    value=>console.log(value),
    ()=>{console.log("Undefined for negative numbers!")}
);
console.log("After")

// Will output:
// Before
// After
// 34
```

As you can see, we chain to the invocation, the `then` method, and pass to it the two functions for success and failure (`resolve` and `reject` in our code). Just like before, we get the same output. Now, this may seem more verbose (it is), but the benefits greatly outweigh the extra typing. Promises are chainable, meaning that for successful operations, you can return a new promise and, that way, have a sequential operation. Here is an example:

```
MyFunction()
    .then(()=>{ return new Promise(...)}, ()=>{...})
    .then(()=>{ return new Promise(...)}, ()=>{...})
    .then(()=>{ return new Promise(...)}, ()=>{...})
    .then(()=>{ return new Promise(...)}, ()=>{...})
    .catch(err=>{...})
```

There are other methods exposed by the `Promise` constructor, such as `.all`, but I will refer you to the documentation to dig deeper into the possibilities and syntax (`https://developer.mozilla.org/en-US/docs/Web/JavaScript/Reference/Global_Objects/Promise`). Still, quite verbose. Lucky for us, JavaScript provides us with a simplified syntax to handle promises, **async/await,** and think of them as a way to code in a more "traditional" way. This only applies to the invocation of promised functions and can only be used in functions.

To see this as an example, let's imagine that we have three functions that return promises, named MyFuncA, MyFuncB, and MyFuncC (yes, I know, not the greatest names). Each one returns, in case of success, *one single value* (this is a condition). These are then used within MyProcessFunction with the new syntax. Here is the declaration:

```
async function myProcessFunction() {              //1
    try {                                         //2
        let    a = await MyFuncA(),               //3
               b = await MyFuncB(),
               c = await MyFuncC()
        console.log(a + b + c)                    //4
    } catch {
        console.log("Error")
    }
}


// Invoke the function normally
MyProcessFunction()                               //5
```

We start by declaring our function with the async keyword (line //1). This signals to the interpreter that we will use the await syntax inside our function. One condition is to wrap the code in a try... catch block. Then, we can use the await keyword in front of the invocation of each promised function call, as in line //3. By line //4, we are certain that each variable has received a value. Certainly, this approach is easier to follow and read.

Let's investigate the equivalences for the line:

```
let a=await MyFuncA()
```

This will match the *thenable* (using .then) syntax:

```
let a;
MyFuncA()
    .then(result=>{ a=result; })
```

However, the problem with this last syntax is that we need to make sure that all the variables a, b, and c have values before we can run line //4, console.log(a+b+c), which would mean chaining the invocations like this:

```
let a,b,c;
MyFuncA()
    .then(result=>{ a=result; return MyFuncB()})
    .then(result=>{ b=result; return MyFuncC()})
    .then(result=>{ c=result; console.log(a+b+c)})
```

This format is harder to follow and certainly more verbose. For these cases, the `async/await` syntax is preferred.

The use of promises is great for wrapping lengthy or uncertain operations and integrating with other patterns that we have seen (façade, decorator, etc.). It is an important pattern to keep in mind that we will use extensively in our applications.

Summary

In this chapter, we have seen principles for software development and important design patterns, with examples in plain JavaScript and, when appropriate, hinted at implementations with Vue 3. These patterns can be hard to grasp the first time you see them, but we will use them and return to them in the rest of the book so that this chapter will work as a reference. This will give you a better idea of when and how to apply different patterns according to the needs of your application.

In the next chapter, we will start to implement a project from scratch and will set the foundations for the applications we will build in the rest of the book. As we move forward, we will reference these patterns to help you consolidate their application.

Review questions

- What is the difference between a principle and a pattern?
- Why is the singleton pattern so important?
- How can you manage dependencies?
- What patterns make reactivity possible?
- Do patterns intertwine? Why? Can you give an example?
- What is asynchronous programming, and why is it so important?
- Can you think of use cases for *promised* functions?

3

Setting Up a Working Project

In the previous chapters, we laid the theoretical foundation for designing a web application in JavaScript using the *Vue 3 framework*. However, so far, we have not really gotten into a real project. That is what this chapter is about. We will use the new set of tools that comes along with Vue 3 to create a project from scratch and prepare a template that we will use in other projects. As is custom, this initial project for a web application is to build a *To-Do list* (the equivalent of *Hello World*). As we progress with the introduction of each new concept, we will over-engineer the application to turn it into something much more useful, or at the very least, more interesting to look upon.

Some of the practical skills we will learn here are as follows:

- Setting up your working environment and **integrated development environment** (IDE)
- Using the new command-line tools and the new **Vite** bundler to scaffold our application
- Modifying the basic template and folder structure to accommodate *best practices* and advanced architecture *design patterns*
- Integrating out-of-the-box **CSS frameworks** into our application
- Configuring the Vite bundler to accommodate our needs

Unlike in previous chapters, this one will be mostly practical, and there will be references to the official documentation for each element of the ecosystem, as these change from time to time. You don't need to memorize the steps, as starting up a project from scratch is not so common for large projects and the tools to scaffold them evolve. Let's start.

Technical requirements

To follow the practical steps in this chapter, you will need the following:

- A computer running **Windows**, **Linux**, or **macOS** with a 64-bit architecture. I will be using **Ubuntu 22.04**, but these tools are cross-platform, and the steps translate between OSs (when something is different, I will point it out).

- **Node.js 16.16.0 LTS** with **npm (node package manager)** installed. You can find the steps to install Node.js in the official documentation, at `https://nodejs.org/`. The building tools work on top of Node.js, so without this, you can't go very far. Node.js is a JavaScript version adapted to run on servers and in systems "outside" the browser, something that makes it very, very handy and powerful. Most of today's bundlers for web development use Node.js in one way or another, if not at least for the great convenience it provides.

- A **text editor** that works with plain text, in UTF-8 format, preferably an IDE. For this tool, there is no shortage of options to choose from. In theory, you can go without an IDE, but I highly recommend that you get one, if not for anything else than the code assistance they provide (also known as **IntelliSense**, **code completion**, etc.). Here are some of the most popular options:

  - **Visual Studio Code** (free): An excellent and very popular option among developers made by Microsoft, which provides good support to Vue 3 through the `Volar` plugin. The official site is `https://code.visualstudio.com/`, and in this book, we will be using this editor as the recommended IDE to work with Vue and Vite.

  - **Sublime Text** (free trial/paid): This is another popular option, especially among macOS users. The official site is `https://www.sublimetext.com/`.

  - **Jetbrains WebStorm** (free trial, paid): The official site is `https://www.jetbrains.com/webstorm/`.

  - **Komodo** IDE (free): The official site is `https://www.activestate.com/products/komodo-ide/`.

  - **NetBeans** IDE (free): The official site is `https://netbeans.apache.org/`.

- A **console** or **terminal emulator**. Users of Linux and macOS will be most familiar with this concept. Windows users can use **Command Prompt**, an integrated terminal on some IDEs, or install a **Windows Terminal** from the Microsoft Store.

- A modern web browser, either based on the Chromium engine (Google Chrome, Microsoft Edge, Opera, Brave, Vivaldi, etc.) or Mozilla Firefox.

With these installed, we are ready to follow the examples and basic projects. However, I would recommend that you also install **Git**, for code versioning control. We will use it later in this book, in *Chapter 9, Testing and Source Control*. In modern development, it is hard to imagine working on a project without some tool to keep track of code changes and version control. Git has become the industry standard. You can install it following the documentation from the official website at `https://git-scm.com/`.

The code files of this chapter can be found on GitHub here: `https://github.com/PacktPublishing/Vue.js-3-Design-Patterns-and-Best-Practices/tree/main/Chapter03`.

Check out the following video to see the Code in Action: `https://packt.link/CmuO9`

Now, with our tools in place, we are ready to start our first project in Vue 3.

Project setup and tools

We will create a new project using **Vite** as our bundler, directly from the command line. Open a terminal window in the directory where you will place our project, and follow these steps:

1. Type the following command:

    ```
    $ npm create vite@latest
    ```

2. If you get a prompt to install additional packages, type Y (yes).

3. Next, you will be prompted to enter project information in the following order:

    A. **Project name**: This will be used to identify the project, and to create a new folder to place it. If you want the project to be installed in the current folder, enter a period (.) as the name.

    B. **Package name**: This name will be used internally for the package configuration. For this example, enter `chapter-3` (or any name of your choosing). This option may not show if you entered or accepted a project name or accepted the default name suggested. If you entered a period (.) as the name to create the project in the current directory, then this option will be mandatory.

    C. **Select framework**: Here, the assistant will display a menu with options. Select `vue` with the arrow keys and press *Enter*.

    D. **Select variant:** Just like before, use the arrow keys and select JavaScript (or TypeScript, but we will use plain JavaScript throughout this book).

Next, you will see how the assistant downloads additional content based on your selections and scaffolds the project. It will create a directory structure with multiple files. However, if we intend to run the project, we soon will discover that it just doesn't work. That is because the scaffolding does not install dependencies, only the skeleton. So, there is still one more step to do, and that is to install the dependencies using npm. In the terminal, enter the following command and hit *Enter* (if you installed in the current directory; if not, first enter the directory just created):

```
$ npm install
```

The package manager will download and install the dependencies for our project and place them in a new directory named `node_modules`. As you can guess already, our development environment for **Vue** with **Vite** is a **Node.js** project.

With the dependencies in place, now is the time to run the project and see what the scaffolding tool prepared for us. In the terminal, enter the following command:

```
$ npm run dev
```

What happens next may be quite fast. Vite will parse your project files and launch a developer server on your machine with a web address that you can use in your browser. You will see something like this in your terminal:

Figure 3.1 - The result of running the development server with Vite

The most important information here is `localhost` and the port where your project website is being served. The milliseconds shown there are just to let you know how fast Vite is to get you up and running (bragging rights, if you ask me). Next, to see the results of our labor so far, open the **local address** in your web browser, and you should be welcomed by a website looking something like the following screen:

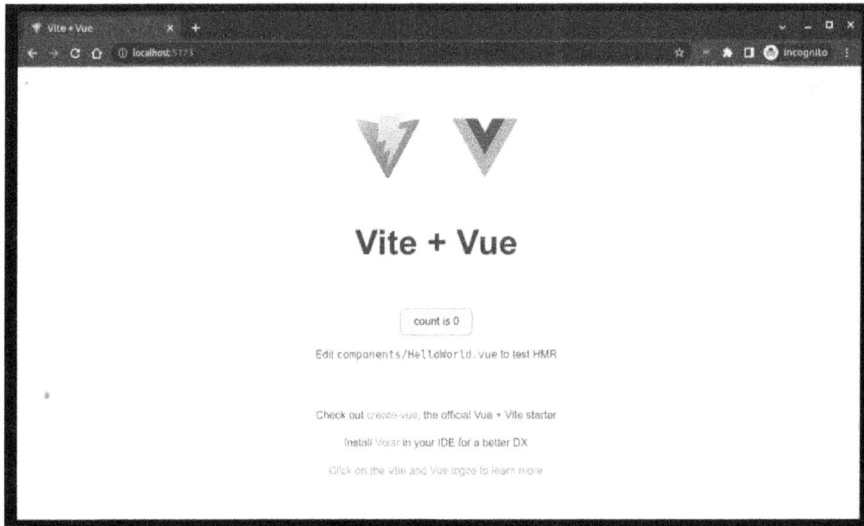

Figure 3.2: Basic Vite + Vue project in the browser

This website is fully functional as it is, if not very productive. To test that Vue 3 is working, click the button in the middle of the screen and you'll see how the counter increases with each click. This is reactivity in action! Moreover, Vite offers us a development server with live updates and **Hot Module Replacement** (**HMR**), which means that as soon as we make changes in our code and save the files, the website will update automatically. In practice, it is very common when developing user interfaces to keep this self-updating site open in the browser to preview our work, and in some cases, several browsers at the same time. Very neat!

We have advanced in our journey, but we are far from over. The scaffolded site is nothing more than a starting point. We will modify it to better serve our purposes and will create a simple To-Do application in the rest of the chapter.

In the next section, we will take a closer look at the structure and organization of our starting project.

Folder structure and modifications

In *Chapter 1, The Vue 3 Framework*, we mentioned that frameworks prescribe some structure for your application. Vue 3 is not an exception, but the conventions used in the directory structure are minimal when compared to other frameworks. If you open the directory where you installed the project in the Files Explorer (either from your OS or in your IDE), you will find a structure like this one:

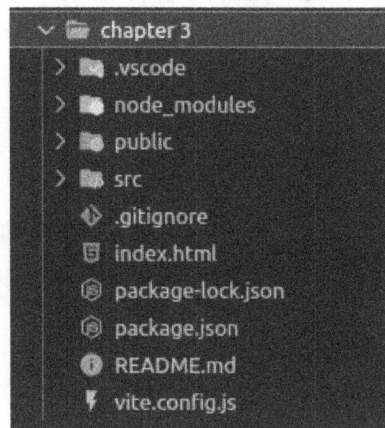

Figure 3.3: Project structure in Visual Code

The `.vscode` folder was created by the IDE, and `node_modules` was created by npm to allocate the dependencies. We will ignore them, as we don't need to worry or work with them. Starting from the top, let's review what each directory is:

- `public`

    This folder contains a directory structure and files that will not be processed by the bundler and will be copied directly into the final website. You can freely place your own static content here. This is where you will place your images, web fonts, third-party CSS libraries, icons, and so on. As a general rule, files here are those that will never be referenced by your code, such as `manifest.json`, `favicon.ico`, `robots.txt` and so on.

- `src`

  Here is where we will place our JavaScript, dynamic CSS, components, and so on. If we expand this folder, we will find that the scaffolding tool has already created a minimal structure with the following:

  - An `assets` folder with an SVG file. In this folder, we can include files that will be manipulated either by the code or by the bundler. You can import them directly into your code, and the bundler will take care of mapping them properly when serving them on a web server.

  - A `components` folder, where we will place our **single-file components** (**SFCs**), with the `.vue` extension. We can create the directory structure here as we please. The scaffolding tool has been placed inside a `HelloWorld.vue` component for us.

  - An `App.vue` file. This is the main component of our application and the root component of our hierarchy. It is a convention to call it this way.

  - A `main.js` file, which is the starting point of our application. It is in charge of loading the initial dependencies, the main component (`App.vue`), creating the Vue 3 application with all extras (plugins, global directives, and components), and launching and mounting the application to the web page.

  - A `styles.css` file, which is a global stylesheet that will apply to our entire application. Previous versions of the scaffolding tool used to place it in the `assets` folder, but now it has moved to the `src/` root giving it a more predominant place. This file, when imported into the `main.js` file, will be parsed and bundled with our JavaScript.

It is now time to investigate the files in the project root, in the same order as they appear:

- `.gitignore` is a file that controls what is excluded from the Git source control. We will see Git in *Chapter 9, Testing and Source Control*.

- `index.html` is the main file and the starting point for our web application. The bundler will start accessing and processing other files in the order they appear, starting with `index.html`. You can modify it to fit your needs, as the generated file is quite basic. Notice how towards the end of the `body` tag, the scaffolding tool included a `script` tag to load our `main.js` file. This file is the one that creates our Vue application. Unlike other bundlers that automatically generate this file and then inject it into `index.html`, Vite requires that you have this imported explicitly. Among other advantages, this gives you control of when the Vue application will be loaded inside your web page.

- `package-lock.json` is used by npm to manage the dependencies in `node_modules`. Ignore it.

- `package.json` is very important. This file defines the project, keeps track of your development and production dependencies, and provides some nice features such as the automation of some tasks by simple commands. Of interest at this time is the `scripts` section, which defines simple aliases for commands. We can run these from the command line by typing `npm run <script name>`. The scaffolding tool already prepared three Vite commands for us:

- `npm run dev`: This will launch the website in developer mode, with a local server and live reload.

- `npm run build`: This will bundle our code and optimize it to create a production-ready version.

- `npm run preview`: This is a middle point between the previous two. It will allow you to see locally the built production-ready version. This may sound confusing until you consider that while in development, the addresses and resources that your application access, as well as the public URL, may be different than those in production. This option lets you run the application locally, but still reference and use those production endpoints and resources. It is a good practice to run a "preview" before you deploy your application.

- `vite.config.js` is the configuration file that governs how Vite will behave during development and when bundling for production. We will see some of the most important or common options later in this chapter.

Now that we have a clearer view of what was given to us by the Vite scaffolding tool, it is time to start building our sample application. Before we dig deep into the code, there are a couple more items we need to address: how to integrate third-party stylesheets and CSS frameworks, and some Vite configurations that will make our life easier.

Integration with CSS frameworks

If we remember the last three principles discussed in *Chapter 2, Software Design Principles and Patterns*, (*don't repeat yourself*, *keep it clean*, and *code for the next*), reinventing the wheel in matters of visual appearance and graphic language is something not desirable in most cases. The web has an ever-growing collection of CSS frameworks and libraries that we can easily incorporate into our applications. From the old popular Bootstrap to atomic design, to utility classes such as Tailwind, and passing by graphics languages such as Material Design and skeuomorphism, the spectrum of options is huge. Vue has a good number of component libraries already implementing some of these libraries, which you can find in the npm repositories. Using these, you'll be restricted to learning about and applying the conventions applied by the designer, which, in some cases, may set in stone how you can build your user interface. Typical examples of these are the use of **Vue-material** (and others) that adheres to Google's Material Design specifications or the incorporation of web fonts and icon fonts. It is impossible to discuss each one, but here are guidelines and some examples of how to incorporate these libraries into your project:

1. Place the static assets provided by the framework or library in the `public` folder, following their required structure, and respecting whatever tree structure is needed.

2. Include the dependencies for the CSS framework or libraries in your `index.html` file, following their instructions. Often, this will imply importing stylesheets and JavaScript files in the `head` section or the `body` tag. In either case, make sure these are placed before the loading of our application (the `script` tag that references our `main.js` file).

3. If the framework or library needs to be instantiated, do so before we mount our application. You can do this directly in `index.html` in a `script` tag, in `main.js`, or in another module.

4. Use the classes (and JavaScript functions) in your component's template section normally, as you would in plain HTML using these libraries. Some frameworks create JavaScript global objects attached to the `window` object, so you can access them directly in your component's `script` section. If this is not the case, consider encapsulating the functionality to use it in your application, using a design pattern such as a *singleton*, *proxy*, or *decorator* patterns.

Now let's put these simple instructions to work by applying them to our example project. We will incorporate a CSS-only framework (meaning that it doesn't use additional JavaScript), and font icons to include basic iconography. In a production build, we should remove unused CSS rules. Some CSS frameworks provide this feature out of the box, such as Tailwind (`https://tailwindcss.com/`). However, this topic is beyond the scope of this book but is worth researching online.

The w3.css framework

The website `w3school.com` offers a free CSS-only framework that is partially based on the Material Design language, made popular by Google, and used in many mobile applications. It offers many utility classes that you can implement, license-free, in your applications. You can find out more on the official website: `https://www.w3schools.com/w3css/`.

We will follow the guidelines mentioned before, so let's follow the steps:

1. Download the `w3.css` file from `https://www.w3schools.com/w3css/w3css_downloads.asp` and place it in a new folder named `css` in the `public` directory. When you are done, it should look like this:

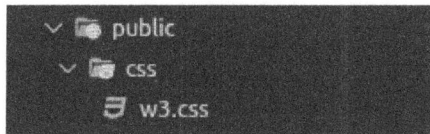

Figure 3.4 - Location for the w3.css file

2. Modify `index.html` at the root of our project by adding the reference to the `w3.css` file using a `link` tag like this:

```
<link rel="stylesheet" href="/css/w3.css">
```

With that inclusion, the classes defined in the CSS file are now ready to be used in our component's templates. Also, to avoid unwelcome styles from the scaffolding of a project, remember to clear the `styles.css` file provided by the installer. If we now run the development server with `npm run dev`, we will see that the appearance of the website has changed slightly, as the new stylesheet has been successfully applied. The next step now is to add an *icon font*.

FontAwesome is just awesome

One of the ways that developers save resources when dealing with a multitude of icons is with the use of **font icons**. These are font files that, instead of characters, display icons. This concept is not new, but it has a lot of applications in web development. Using fonts for icons, as opposed to other techniques (such as CSS sprite sheets, for example) has plenty of benefits. One of the most relevant is that these icons are subject to the same manipulation as regular fonts, so we can easily alter their size, color, and so on, and keep them in coordination with the rest of the text. Not all is joy and happiness with this approach, since now, the major trade-off is that these icons display only one or two colors at most and have to be rather simple by necessity.

FontAwesome is a website that offers **icon fonts** to use in our applications, both web and desktop. It has been doing this for years and has some of the best icon collections out there. We can download and use its free tier for our project. Let's follow again the guidelines to implement them in our project:

1. Download the fonts *for the web* from `https://fontawesome.com/download`. It will download a rather large ZIP file with all the different alternatives.

2. From within the ZIP file, copy the `css/` and `webfonts/` directories as they are, to our `public/` folder. We will not use everything in this folder in our projects here, so you can delete what we don't use later.

3. Edit the `index.html` file to add the stylesheets that we will use. These CSS files will automatically load the icon fonts from the `/webfonts/` folder:

```
<link rel="stylesheet" href="/css/fontawesome.min.css"
>
<link rel="stylesheet" href="/css/solid.min.css">
<link rel="stylesheet" href="/css/brands.min.css">
```

And that is all we need to do to include FontAwesome in our project. There are other alternatives that have encapsulated the fonts into Vue components, and even the website provides a Vue implementation. However, for our purposes in this book, we will use the *direct* approach. If we open the icons section of the site, we can browse and search all the available icons. You can restrict the search to "solid" and "brands" since those are the ones we have included in our project. For example, if you want to display the Vue icon using FontAwesome, we can include the following in our template:

```
<i class="fa-brands fa-vuejs"></i>
```

These classes make all the magic happen in any empty element, but for tradition and convenience, we always use the i tag. Moreover, you do not even need to type them. Once you locate the icon you want to use, the website offers this neat feature to "click and copy" the code. The previous line came from here:

Figure 3.5 - FontAwesome icon page

Let's keep in mind that including a large library of icons when only using a few of them will impact performance. For production builds, ensure that you only include the icons you will use in your application by creating icon fonts only with the necessary icons. For the purposes of our book and during development, we can skip this practice.

With a nice stylesheet and some good icon fonts, we are almost ready to start our coding. There is just one more thing to do and it is to include a few extra options in our Vite configuration.

Vite configuration options

The `vite.config.js` file exports the configuration that Vite will use for development and also for production. Vite was meant to be functional for many different frameworks and not only for Vue 3, even though it is the official bundler for it. When we open the file, we notice that Vue is a plugin for Vite. Internally, Vite uses **Rollup.js** (`https://www.rollupjs.org/`) and **esbuild** (`https://esbuild.github.io/`), for development and production build, respectively. This means that we can pass options to Vite, but also have even more fine-grained control over some edge cases by passing arguments to these two underlining tools. Additionally, you can pass different configurations for each processing mode (development and production), so we are not left without options here.

We will see some specific configurations for deployment in *Chapter 10, Deploying Your Application*, but for now, we will focus only on the development part with a few additions to keep us from typing too much and repeating ourselves in the code.

Open the `vite.config.js` file and add the following import:

```
import path from "path"
```

Yes, the path import is not JavaScript, but Node.js, and we can do this because this file is read and executed in a Node.js context. It will never reach the browser or any JavaScript context.

Modify the export configuration so it looks like this:

```
export default defineConfig({
plugins: [vue()],
  resolve:{
    alias:{
      "@components":
          path.resolve(__dirname, "src", "components")
    }
  }
})
```

In these lines, we are indicating an alias named `@components` matched to the project path `/src/components`. This way, when we are importing components, we can avoid writing relative or full paths, and just reference the imports inside components in this manner:

```
import MyComponent from "@components/MyComponent.vue"
```

Having aliases for paths is a nice feature to have for a developer experience. Paths to components can get quite long in a large project, and code reorganization does happen from time to time, making maintenance yet another point of possible disruption. Having an alias defined gives us more flexibility by making changes only in one place (*principle: Don't repeat yourself*).

You can find a complete reference of the Vite configuration file at `https://vitejs.dev/config`. Vite offers a short list of official plugins (such as for Vue) at `https://vitejs.dev/plugins/`, but the community has provided a fair share of plugins to cover many scenarios at `https://github.com/vitejs/awesome-vite#plugins`. These can be installed and imported into our configuration file when and if needed.

At this point, we have enough preparation done so we can move ahead and finally create our simple To-Do app.

The To-Do app

Our example application will build on the scaffolding files of a basic application. It will provide us with an input element to enter our to-do items and will display the list of tasks pending and completed. The purpose of this exercise is as follows:

- Develop the application with live updates
- Create a component, with reactive elements in the `script setup` syntax
- Apply styles and icon fonts from third-party libraries

When we are done, we will have a simple website that should look like this (the to-do items have been added as an example):

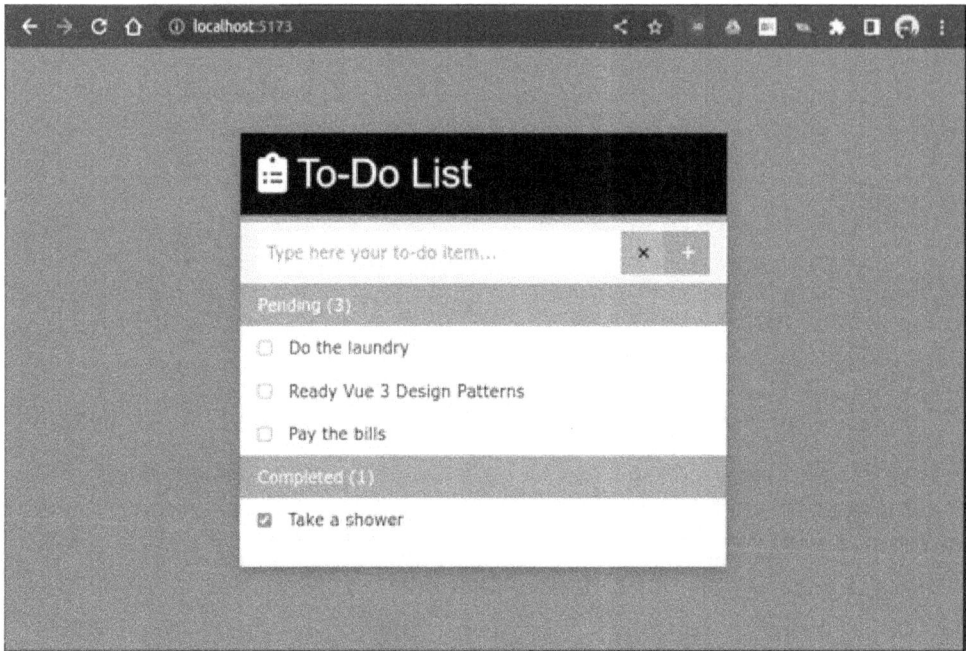

Figure 3.6 - The final result of our To-Do List application with styles applied

For the purpose of this exercise, we will develop the entire To-Do application in one single component, which we will import into our main component (App.vue). This, of course, is purposely breaking some of the principles that we saw in *Chapter 2, Software Design Principles and Design Patterns*. In *Chapter 4, User Interface Composition with Components*, we will take this product and "make it right" with multiple components.

In the application, the user will do the following:

1. Type a short description and press the *Enter* key or click the plus sign to enter it as a task.
2. The system will display the pending and completed tasks in separate lists, displaying how many are in each group.
3. The user can click on any task to mark if it has been done or undone, and the application will move it to the corresponding group.

Knowing how the application has to work, let's move onto the code.

App.vue

This is our main component. In the starter application, we need to remove the content from each section and change it to the following (we'll explain next what each part does):

```
<script setup>
    import ToDos from "@components/ToDos.vue"
</script>
```

In the `script` section, we only need to import a component named ToDos (we will create this file next). Notice how we are using already an alias for the path (@components). Our main component will not handle any other data or functionality, and we are using it only as a wrapper to control the layout of this application. With that in mind, our template now will look like this:

```
<template>
  <div class="app w3-blue-gray">
    <ToDos />
  </div>
</template>
```

We declare a `div` element with a private class (.app) that we will define in the `style` section. We also applied one of the styles we imported from W3.css to give our application a background color. Inside our `div` element, we place our ToDos component. Notice that we are using the same name as imported into our `script` section in Pascal case. We can use this notation, or the HTML kebab-case equivalent, `<to-dos />`, (lowercase words separated by hyphens). However, the recommendation is to always use Pascal case in our templates, with multiple words, to avoid conflicts with HTML components, present or future. This name will be transformed into kebab case in the final HTML.

Next, we will define the style, using the CSS `flex` layout to center our component in the center of the screen:

```
<style scoped>
.app {
  display: flex;
  justify-content: center;
  width: 100vw;
  min-height: 100vh;
  padding: 5rem;
}
</style>
```

With our main component in place, now let's create our ToDos component in the /src/components directory, properly named ToDos.vue.

ToDos.vue

In this component, we will place all the logic of this simple application. We will need the following reactive variables:

- A variable to capture the text from the input, and create our task

- An array where we will host our task objects with the following fields: a unique ID, a description, and a Boolean value to indicate whether it is complete or not

- A filtering function or computed property (or properties) to display only the tasks that are completed

With the preceding requirements, let's populate our `script` section with the following code:

```
import { ref, computed } from "vue"                          //1
const                                                        //2
    _todo_text = ref(""),
    _todo_list = ref([]),
    _pending = computed(() => {                               //3
        return _todo_list.value.filter(item =>
                                        !item.checked)
    }),
    _done = computed(() => {                                  //4
        return _todo_list.value.filter(item =>
                                        item.checked)
    })
function clearToDo() {_todo_text.value = ""}                  //5
function addToDo() {                                          //6
    if (_todo_text.value && _todo_text.value !== "") {
        _todo_list.value.push({id:  new Date().valueOf(),
        text: _todo_text.value, checked: false})
        clearToDo()
    }
}
```

We start by importing the `ref` and `computed` constructors from Vue in line `//1`, since this is all that we will need in this application from the framework. In line `//2`, we start declaring two constants to point to reactive values: `_todo_text`, which will host our user's task description in the input element, and `_todo_list`, which will be the array of tasks (to-do items). In lines `//3` and `//4`, we declare two `computed` properties named `_pending` and `_done`. The first one will have a reactive array of all the to-do items that are incomplete, and the second one all those marked as completed. Notice that by using a `computed` property, we only need to keep a single array with all the items. Computed properties are used to get a view segment of the list according to our needs. This is a commonly used pattern for these kinds of circumstances as opposed to, for example, having two arrays for each group and moving items between them.

Finally, we have a helper function in line //5, to reset the value of our item text, and in line //6, we have a simple function that checks the value of the description and creates a task (to-do item) to add to our list. It is important to note that the moment we modify _task_list, all the properties and variables that depend on it will be automatically re-evaluated. Such is the case with the computed properties.

This is all that we will need in our component's logic to achieve the results we want. Now, it is time to create the template with HTML. We will split the code into sections for convenience. The segments that appear highlighted mark those with bindings or interactions with the framework and our code in the script section:

```
<div class="todo-container w3-white w3-card-4">              //1
    <!-- Simple header -->                                  //2
    <div class="w3-container w3-black w3-margin-0
        w3-bottombar w3-border-blue">
        <h1>
            <i class="fa-solid fa-clipboard-list"></i>
            To-Do List
        </h1>
</div>
```

The template of our component starts in line //1 by defining a wrapper element with some styles. Then, in line //2, we place a simple header with styles and an icon font. Notice how we are using CSS classes from the **W3 CSS framework**, at the same time as our own scoped styles. The next lines in the code will focus on capturing the user input:

```
<!-- User input -->                                         //3
<div class="w3-container flex-container w3-light-gray w3-padding">
    <input class="w3-input w3-border-0" type="text"
            autofocus
            v-model="_todo_text"
            @keyup.enter="addToDo()"
            placeholder="Type here your to-do item...">
    <button class="w3-button w3-gray" @click="clearToDo()">
        <i class="fa-solid fa-times"></i>
    </button>
    <button class="w3-button w3-blue" @click="addToDo()">
        <i class="fa-solid fa-plus"></i>
    </button>
</div>
```

The interactivity with the user starts in the section in line //3, where we define an input element and attach our _todo_text reactive variable with the v-model directive. From this time on, anything that the user types into our input box will be the value of our variable in the code. Just for convenience, we are also capturing the *Enter* key with the following attribute:

```
@keyup.enter="addToDo()"
```

This will trigger the addToDo function from our script. We will add the same in the plus button next to the input field, also on the click event:

```
@click="addToDo()"
```

This gives us two ways to enter our descriptions as tasks for our to-do list, using multiple events linked to the same function. The following code now focuses on displaying the input data:

```
<!-- List of pending items -->                                         //4
<div class="w3-padding w3-blue">Pending ({{ _pending.length }})
</div>
<div class="w3-padding" v-for="todo in _pending" :key="todo.id">
    <label>
        <input type="checkbox" v-model="todo.checked">
        <span class="w3-margin-left">
            {{ todo.text }}
        </span>
    </label>
</div>
<div class="w3-padding" v-show="_pending.length == 0">No tasks
</div>
<!-- List of completed tasks -->                                        //5
<div class="w3-padding w3-blue">Completed ({{ _done.length }})
</div>
<div class="w3-padding" v-for="todo in _done" :key="todo.id">
    <label>
        <input type="checkbox" v-model="todo.checked">
        <span class="w3-margin-left">
            {{ todo.text }}
        </span>
    </label>
</div>
<div class="w3-padding" v-show="_done.length == 0">                      //6
  No tasks
</div>
</div>
```

To display our task list, we have two almost identical blocks of code, starting on lines //4 and //5 – one for the pending tasks and the other for the completed ones. We will only focus on the first block (starting on line //4) since the behavior of these blocks is almost the same. In the first div element, we create a small header that displays the number of items on the _pending array, by interpolating its length. We do this with the following line:

```
Pending ({{ _pending.length }})
```

Notice how we can access the array attributes directly inside the double curly brackets, without the use of the .value attribute. While in our JavaScript code, we should write this as _pending. value.length, when we are using interpolation in our HTML, Vue is smart enough to identify the reactive variable in our template section and access the value directly. This is true for computed properties as well as reactive variables created with ref().

In the next div element, we create a list with a v-for/:key directive that will iterate over our _pending array and create a copy of the element for each item. Inside each one, we can now reference each item with the name todo, which we declared in the v-for directive. Next, we wrap an input checkbox and a span inside a label element and bind the todo.checked property (Boolean) to the input with v-model. Vue will take care of assigning a true or false value depending on the state of the checkbox. When that happens, it will also trigger the recalculation of the computed properties and we will see how just by checking/unchecking an item, it moves between groups (pending and completed) and also updates the total number of each block. We also have a span element to display the text of the task.

Finally, for the cases when a list group is empty, we also have a div element that will be visible only when that list is empty in line //6 (_pending.length==0).

As mentioned before, the part that displays our "done" list works in the same way, applying the same logic.

Our scoped styles will be quite small in this case, as we only need a couple of extra settings since most of the heavy lifting has been done using the w3.css library. Inside our style section, add the following:

```
.todo-container {max-width: 100%; min-width: 30rem;}
label {cursor: pointer; display: flex;}
```

The todo-container class limits the maximum and minimum width of our component, and we also modify the label element to display its children using the flex layout.

To see the application in action, save all the changes and start the Vite development server with the following command in the terminal:

```
$ npm run dev
```

Once Vite is ready, open the address in the web browser just as we did before. If all is well, you should see our To-Do list working as expected. If not, check with the source code in the repository to make sure that your typed code matches the full example.

A quick critique of our To-Do application

The application we just made is working and is a bit more advanced than a simple `Hello World` or a counter button. However, we have not applied all the best practices and patterns that we should or could. This is done on purpose, as a learning exercise. Sometimes, to know how to build something right, we first need to build it to work as is. In general, all engineering practices understand that there is an iterative refinement process that provides learning and sophistication with each interaction. Once we build our first prototype, it is time to take a step back and do a sincere critique of it, thinking about how we can improve it and do things better. In this case, here is our critique:

- We have duplication of code in our template, as the `_pending` and `_done` computed properties are basically the same, with a minor difference based on the value of a variable.

- We are not leveraging the power of components, as everything is built in a single component.

- Our component is also creating our models (the To-Do items), so our business logic is tied to our component.

- We have done very little in terms of input sanitization and control. It is foreseeable that some code, even equal inputs, will break our application.

- Our to-do list is volatile. A refresh of the page will wipe clean our list.

- Our task only accommodates two states (done and pending). What if we want to have a third state or more? For example, in progress, waiting, or next in line?

- The current design does not provide a way to edit or delete a task once it has been created.

- We can only manage one list of items at a time.

As we move forward, we will improve our application and apply principles and patterns to make this a more resilient and useful application. In the next chapter, we will look at how to compose a web application with web components in a more approachable way.

Summary

In this chapter, we have started to create applications using real-life tools, from IDEs to command-line tools, to scaffold, preview, and build our application. We have also created a simple To-Do application and learned how we can integrate third-party CSS libraries and icon fonts into our application and defined some general guidelines to incorporate others. We also took a critical approach to our simple application as a step to improve its functionality and, at the same time, our skills. In the next chapter, we will look into how to better organize our code and create a component hierarchy to create our user interfaces.

Review questions

- What are the requirements to develop a Vue 3 application with Vite?

- Is it possible to integrate third-party libraries and frameworks with Vue 3?

- What are some steps to integrate a CSS-only library with a Vue application?

- Is creating an application inside a single component a good idea? Why yes or no? Can you think of scenarios when a single-component application is the right fit? How about a scenario when it is not?

- Why is software development an iterative refinement process?

4

User Interface Composition with Components

In this chapter, we will take a closer look at how to compose user interfaces with components. While we could just create our entire web page with just one component, as we did with our initial *To-Do list* application in *Chapter 3, Setting Up a Working Project*, this approach is not a good practice save for simple applications, partial migrations of functionality in existing web applications, or some edge cases when there could be no other option. Components are central to Vue's approach to building interfaces.

In this chapter, we will do the following:

- Learn how to compose user interfaces with a hierarchy of components
- Learn different ways in which components interact and communicate with each other
- Look into special and custom components
- Create an example plugin applying design patterns
- Re-write our to-do application using our plugin and component composition

This chapter will introduce core and advanced concepts and give you the tools for building solid web applications with reusable components. In particular, we will apply our knowledge of design patterns from *Chapter 2, Software Design Principles and Patterns*, in the implementation of the code.

> **A note about styles**
>
> To avoid lengthy code listings, we will omit sample icons and styles in the code sample. The complete code, along with the styles and iconography, can be found in this book's GitHub repository at `https://github.com/PacktPublishing/Vue.js-3-Design-Patterns-and-Best-Practices`.

Technical requirements

The requirements to follow this chapter are the same as previously mentioned in *Chapter 3, Setting Up a Working Project.*

Check out the following video to see the Code in Action: `https://packt.link/eqm4l`

The code files of this chapter can be found on GitHub here: `https://github.com/Packt-Publishing/Vue.js-3-Design-Patterns-and-Best-Practices/tree/main/Chapter04`

Page composition with components

To create a user interface, we must have a starting point, be it a crude sketch to a fancy full-fledged design. The graphic design of a web application is beyond the scope of this book, so we will consider that it has been created already. To translate the design into components, we could approach it as a process that answers the following questions:

1. How can we represent the layout and multiple elements with components?

2. How will these components communicate and relate to each other?

3. What dynamic elements will enter or leave the scene, and what events or application states will they be triggered by?

4. What design patterns can we apply that best serve the use case, considering trade-offs?

Vue 3 is specially fit to create dynamic, interactive interfaces. These questions lead us to a repeatable approach for the implementation. So, let's define a general process with well-defined stages, step by step.

Step 1 – identify layouts and user interface elements

This step answers the question: *How can we represent the layout and multiple elements with components?*

We will take the page as a whole and consider what layout would work best, given the design. Should we use columns? Sections? Navigation menus? Islands of content? Are there dialogs or modal windows? A simple approach is to take the design image and mark the sections that may represent components with rectangles, from the outermost down to the singular unit of interaction. Iterate over this *slicing* of the page until you have a comfortable number of components. Considering the new To-Do application design, here is what this step may look like:

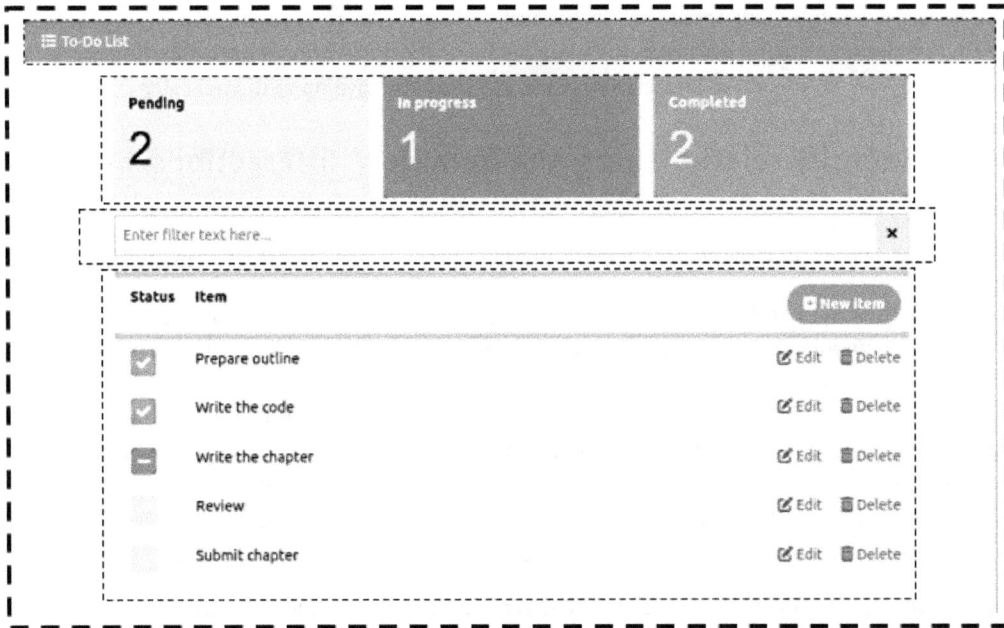

Figure 4.1 – A slicing of the design into components with dashed boxes

Once we've identified the components, we must extract the relationships between them, creating a hierarchy from the topmost root component (usually, this will be our App.vue). New components may appear because of grouping components by context or functionality. This is a good time to name the components. This initial architecture will evolve as we implement design patterns. Following this example, the hierarchy may look something like this:

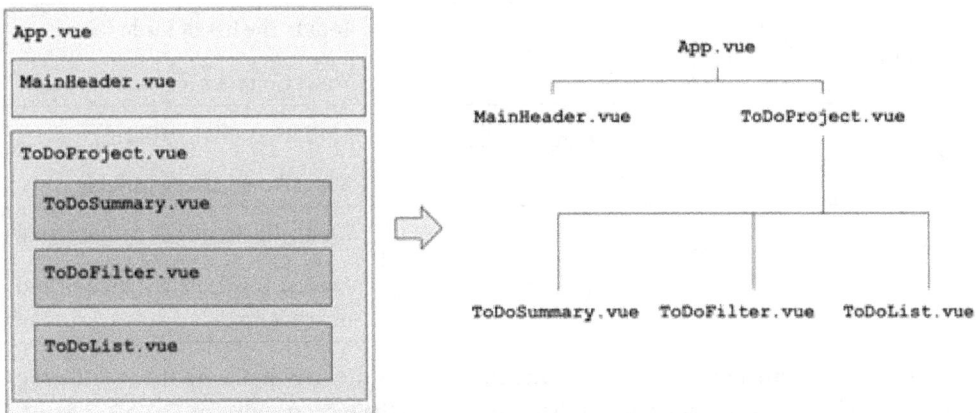

Figure 4.2 – An initial approach to the component's hierarchy

Notice how a new component appeared, `ToDoProject.vue`, from grouping other components. The `App` component usually deals with the main layout of the application and the starting point in the hierarchy. Now, with our initial design in place, it is time to move on to the next step.

Step 2 – identify relationships, the data flow, interactions, and events

This step answers the question: *How will these components communicate and relate to each other?*

In this stage, we need to understand the user's interaction (with a use case, user story, or something else). For each component, we decide what information it will hold (the state), what will pass down to its children, what it needs from its parent, and what events it will trigger. In Vue, components can only relate vertically to one another. Siblings ignore the existence of each other for the best part. If a sibling component needs to share data with another, that data must be hosted by a common third party who can share it with both, usually the parent who has common visibility. There are other solutions for this, such as reactive state management, which we will cover in detail in *Chapter 7, Data Flow Management*. For this chapter, we will settle with the basic relationship functionality.

There are many ways to document this information: scribbled notes in the hierarchy tree (see *Figure 4.2*), descriptive formal documentation, UML diagrams (**UML** stands for **Universal Modeling Language**, an iconography representation of software components), and more. For simplicity, let's write down only one segment of the tree in a table format:

Component	Function	State, I/O, events
ToDoProject.vue	Hosts a list of to-do items and coordinates interaction with the user. This component will actively modify the items.	State: The to-do list Events: Open new, edit, and delete modals
ToDoSummary.vue	Displays a summary count of to-do items by state.	Input: The list of to-do items State: Counters for each item state
ToDoFilter.vue	Collects a string to filter the list of to-do items.	Output: A filter string State: An auxiliary variable
ToDoList.vue	Displays the list of to-do items, and the signal operations for each one.	Input: The to-do list, a filter string Events: Toggle item state, edit and delete item

For brevity, I have omitted the components and interactions that will make up the user dialogs. We will see them later in this chapter, but suffice it to say, it is the responsibility of `ToDoProject.vue` to manage the interaction using modal dialogs.

Step 3 – identify user interactivity elements (inputs, dialogs, notifications, and more)

This step answers the question: *What dynamic elements will enter or leave the scene, and what events or application states will they be triggered by?*

In our application, the main CRUD operations (**CRUD** stands for **Create, Read, Update, and Delete** data) involve using modal dialogs presented to the user. As previously mentioned, it is the `ToDoProject.vue` component that controls this interaction as a response to certain events. This process is illustrated in this sequence diagram:

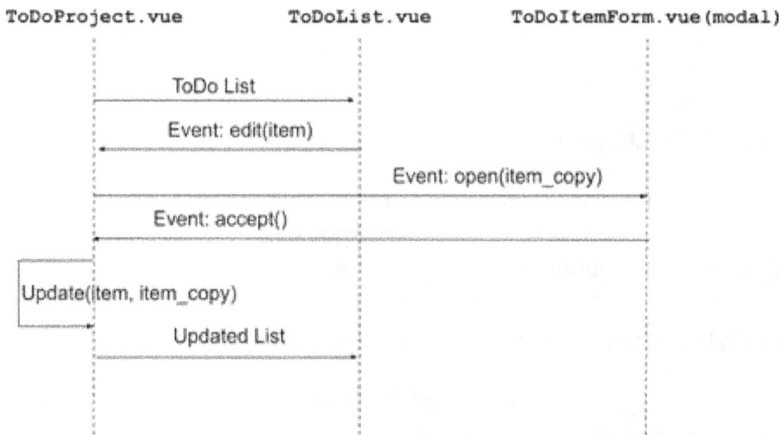

Figure 4.3 – User interaction through modals – edit an item

In this diagram, the `ToDoProject` component shares the to-do list with the `ToDoList` component. When the user triggers the `edit` event, the child component notifies the parent by raising such an event. The parent then makes a copy of the item and opens a modal dialog, passing said copy. When the dialog is accepted, the parent modifies the original item with the changes. Then, Vue's reactivity reflects the state change in the child components.

Often, these interactions help us identify the need for additional components that were not evident in *Step 1*, such as the implementation of design patterns… which is the next step.

Step 4 – identify design patterns and trade-offs

This step answers the question: *What design patterns can we apply that best serve the use case, considering trade-offs?*

Deciding what patterns to use can be a very creative process. There is no silver bullet, and multiple solutions can provide different results. It is common to make several prototypes to test different approaches.

In our new application, we have introduced the concept of modal dialogs to capture user input. Modal dialogs are used when an operation requires a user action or decision to proceed. The user can accept or reject the dialog, and cannot interact with any other part of the application until such a decision is made. Given these conditions, one possible pattern to apply is the **Async Promise** pattern.

In our code, we want to open a modal dialog as a promise that, by definition, will provide us with a `resolve()` (accept) or `reject()` (cancel) function. Moreover, we want to be able to use this solution in multiple projects, and globally in our application. We can create a plugin for this purpose, and use the **dependency injection pattern** to access the modal functionality from any component. These patterns will provide us with the solution we need to make our modal dialog reusable.

At this point, we are almost ready to start implementing the components conceptually. However, to create a more suitable and sturdy application, and implement the aforementioned patterns, we should take a moment to learn more about Vue components.

Components in depth

Components are the building blocks of the framework. In *Chapter 1, The Vue 3 Framework*, we saw how to work with components, declare reactive variables, and more. In this section, we will explore more advanced features and definitions.

Local and global components

When we start our Vue 3 application, we mount the main component (`App.vue`) to an HTML element in the `main.js` file. After that, in the script section of each component, we can import other components to use locally through this command:

```
import MyComponent from "./MyComponent.vue"
```

In this manner, to use `MyComponent` in another component, we need to import it again in such a component. If one component is used continuously in multiple components, this repetitive action breaks the development DRY principle (see *Chapter 2, Software Design Principles and Patterns*). The alternative is to declare the component as **global**, by attaching it directly to our Vue application instead of each component. In the `main.js` file, we can use the `App.component()` method for this use case:

Main.js

```
Import { createApp } from "vue"
import App from './App.vue'
Import MyComponent from "./MyComponent.vue"

createApp(App)
    .component('MyComponent', MyComponent)
    .mount("#app")
```

The `component()` method receives two arguments: a `String` that represents the HTML tag for the component, and an object with the component definition (either imported or in-lined). After registration, it is available to all the components in our application. There are, however, a few drawbacks to using global components:

- The component will be included in the final build, even if never used

- Global registrations obscures the relationship and dependencies between components

- Name collision may occur with locally imported components

The recommendation is to globally register only those components that provide generic functionality and avoid those that are an integral part of a workflow or specific context.

Static, asynchronous, and dynamic imports

So far, all the components we have imported have been *statically* defined with the `import XYZ from "filename"` syntax. Bundlers such as **Vite** include them in a single JavaScript file. This increases the bundle's size and could create delays in the startup of our application as the browser needs to download, parse, and execute the bundle and all its dependencies before any user interaction may take place. This code may include features that are seldom used or accessed. The clear alternative to this is to split our bundle file into multiple smaller files and load them as needed. In this case, we have two approaches – one provided by Vue 3 and another one provided by the newest JavaScript syntax for dynamic imports.

Vue 3 provides a function called `defineAsyncComponent`. This function takes a parameter another function that returns a dynamic import as an argument. Here is an example:

```
import {defineAsyncComponent} from "vue"
const MyComponent = defineAsyncComponent(
                    ()=>import("MyComponent.vue")
              )
```

The use of this function makes it safe to use in most bundlers. An alternative to this syntax is used by Vue Router, which we will see in *Chapter 5, Single-Page Applications*: the `import()` dynamic declaration provided by JavaScript (`https://developer.mozilla.org/en-US/docs/Web/JavaScript/Reference/Operators/import`), This has a very similar syntax:

```
const MyComponent = () => import('./MyComponent.vue')
```

As you can see, this syntax is more succinct. However, it can only be used when defining routes with Vue Router as, internally, the way that Vue 3 and Vue Router handle lazy loading components is different. In the end, both approaches will split the main bundle file into multiple smaller files that will be automatically loaded when needed in our application.

However, `defineAsyncComponent` has some advantages. We can also pass any function that returns a promise that resolves to a component. This allows us to implement logic to control the process dynamically at runtime. Here is an example where we have decided to load one component based on the value of an input:

```
const ExampleComponent=defineAsyncComponent(()=>{
    return new Promise((resolve, reject)=>{
        if(some_input_value_is_true){
            import OneComponent from "OneComponent.vue"
                resolve(OneComponent)
        }else{
            import AnotherComponent from
                "AnotherComponent.vue"
            resolve(AnotherComponent)
        }
    })
})
```

The third syntax for `defineAsyncComponent` is probably the most useful. We can pass an object with attributes as an argument, which provides more control over the loading operation. It has these attributes:

- `loader` (mandatory): It must provide a function that returns a promise that loads the component
- `loadingComponent`: The component to display while the asynchronous component is loading
- `delay`: The number of milliseconds to wait before displaying `loadingComponent`
- `errorComponent`: The component to display if the promise rejects, or if the loading fails for any reason
- `timeout`: The time in milliseconds before considering the operation to have failed and displaying `errorComponent`

Here is an example that uses all these attributes:

```
const HeavyComponent = defineAsyncComponent({
    loader: ()=> import("./HeavyComponent"),
    loadingComponent: SpinnerComponent,
    delay: 200,
    errorComponent: LoadingError,
    timeout: 60000
})
```

While the browser retrieves the component from the `loader` attribute, we display a SpinnerComponent to inform the user that the operation is underway. After 1 minute of waiting, as defined by `timeout`, it will display the LoadingError component automatically.

With this approach, our code is better optimized. Now, let's learn how to receive data and notify other components through events.

Props, events, and the v-model directive

We have seen basic uses for props and events as means of passing data in and out of a component to its parent. But more powerful definitions are possible with multiple syntaxes. Props can be defined in the `script setup` syntax with `defineProps` and any of the following argument formats:

- As an array of strings – for example:

  ```
  const $props=defineProps(['name', 'last_name'])
  ```

- As an object, whose attributes are used as a name, and the value is of the data type – for example,

  ```
  const $props=defineProps({name: String, age: Number})
  ```

  As an object, whose attributes define an object with a type and default value – for example,

  ```
  const $props=defineProps({
      name: { type: String, default: "John"},
      last_name: {type: String, default: "Doe"}
  })
  ```

We need to keep in mind that primitive values are passed to the component by **value** (meaning that changing their value inside the child component will not affect their value in the parent). However, complex data types, such as objects and arrays, are passed as **references**, so changes to their inner keys/values will reflect in the parent.

> **A note on complex types**
>
> When defining props of the `Object` or `Array` type with default values, the default attribute must be a function that returns said object or array. Otherwise, the reference to the object/array will be shared by all the instances of the component.

Events are signals that our child component emits to the parent. This is an example of how to define the events for a component in the `script setup` syntax:

```
const $emit=defineEmits(['eventName'])
```

Unlike props, emits only accept an array of strings declaration. Events can also pass a value to the receiver. Here is an example of the invocation from the aforementioned definition:

```
$emit('eventName', some_value)
```

As you can see, `defineEmits` returns a function that accepts one of the same names provided in the definition array as the first argument. The second argument, `some_value`, is optional.

Custom input controllers

One special application with props and events acting together is to create custom input controllers. In the previous examples, we used the Vue `v-model` directive on basic HTML input elements to capture their value. Props and events that follow a special naming convention allow us to create input components that accept the `v-model` directive. Let's take a look at the following code:

Parent component template

```
<MyComponent v-model="parent_variable"></MyComponent>
```

Now that we have the `MyComponent` in use inside our parent component, let's see how we create the tie in:

MyComponent script setup

```
const $props=defineProps(['modelValue']),
      $emit=defineEmits(['update:modelValue'])
```

We are using the array definition of `Props` for brevity. Notice that the name of the prop is `modelValue`, and the event is `update:modelValue`. This syntax is expected. When the parent assigns a variable with `v-model`, the value will be copied to `modelValue`. When the child emits the `update:modelValue` event, the parent variable's value will be updated. In this way, you can create powerful input controls. But there's more – you can have multiple `v-models`!

Let's consider that `modelValue` is the default when using `v-model`. Vue 3 has introduced a new syntax for this directive so that we can have multiple models. The declaration is very simple. Consider the following child component's declaration:

Child component props and event

```
const
  $props=defineProps(['modelValue', 'title']),
  $emit=defineEmits(['update:modelValue','update:title'])
```

With the preceding props and emits definition, we can now reference both from the parent component as the following example shows:

Parent component template

```
<ChildComponent v-model="varA" v-model:title="varB"></ChildComponent>
```

As we can see, we can attach a modifier to the `v-model:name_of_prop` directive. In the `Child` component, the event's name now has to include the `update:` prefix.

The use of props and events allows a direct data flow to occur between parent and child components. This implies that if data needs to be shared with multiple children, it has to be managed at the parent level. One issue with this restriction appears when the parent needs to pass data not to a child, but to a grandchild or other deeply nested component in the hierarchy tree. That is where the *dependency injection pattern* comes in to save the day. Vue implements this naturally with the `Provide` and `Inject` functions, which we will cover in more detail in the next section.

Dependency injection with Provide and Inject

When data in the parent needs to be available in a deeply nested child, using only props, we would have to "pass" the data between components, even if they don't need it or use it. This issue is called *props drilling*. The same occurs with events traveling in the opposite direction, having to "bubble" upwards. To solve this issue, Vue offers an implementation of the dependency injection pattern with two functions named `Provide` and `Inject`. Using these, the parent or root component *provides* data (either by value or reference, such as an object), that can be *injected* into any of its children down the hierarchy tree. Visually, we can represent this situation as follows:

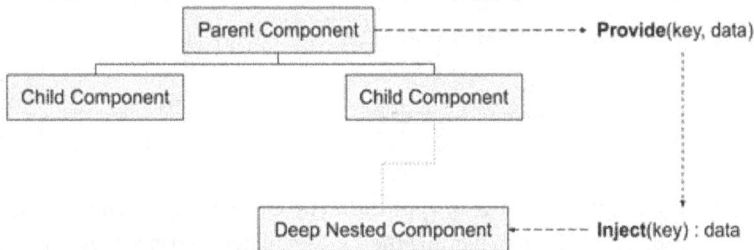

Figure 4.4 – Representation of Provide/Inject

As you can see, the process is very simple, as well as the syntax to implement the pattern:

1. In the parent (root) component, we import the `provide` function from Vue and create a provision with a key (name) and the data to pass along:

    ```
    import {provide} from "vue"
    provide("provision_key_name", data)
    ```

2. In the receiving component, we import the `inject` function and retrieve the data by key (name):

    ```
    import {inject} from "vue"
    const $received_data = inject("provision_key_name")
    ```

We can also provide a resource at the application level in the following manner:

```
const app = createApp({})
app.provide('provision_key_name', data_or_value)
```

In this way, the provision can be injected into any component of our application. It is worth mentioning that we can also provide complex data types, such as arrays, objects, and reactive variables. In the following example, we are providing an object with functions and references to the parent methods:

In the parent/root component

```
import {provide} from "vue"
function logMessage(){console.log("Hi")}
const _provision_data={runLog: logMessage}
provide("service_name", _provision_data)
```

In the child component

```
import {inject} from "vue"
const $service = inject("service_name")
$service.runLog()
```

In this example, we have effectively provided an **application programming interface** (**API**) through an object system-wide. A good practice when naming the "provision key" (service name) is to adhere to a convention that the entire team will understand and follow to identify the functionality, context, and maybe the source of the service provided, and to avoid possible collisions. For example, an injectable service named `Admin.Users.Individual.Profile` is more descriptive than `user_data`. It is up to the team and the developer to define the naming convention (a path-like naming is just a suggestion, not a standard). As mentioned before in this book, once you've decided upon a convention, what matters is that you are consistent throughout the source code. Later in this chapter, we will use this method to create a plugin to display modal dialogs, but before that, we need to see a few more concepts regarding special components and templates.

Special components

The hierarchy of components is very powerful but has limitations. We have seen how we can apply the dependency injection pattern to solve one of those, but there are other cases where we need a bit more flexibility, reusability, or power to share code or templates, or even move a component that's rendering outside the hierarchy.

Slots, slots, and more slots...

Through the use of props, our component can receive JavaScript data. With analog reasoning, it is also possible to pass template fragments (HTML, JSX, and so on) into specific parts of a component's template using placeholders called **slots**. Just like props, they accept several types of syntax. Let's start with the most basic: the *default slot*.

Let's assume we have a component named MyMenuBar that acts as a placeholder for a top menu. We want the parent component to populate the options in the same way that we use a common HTML tag such as header or div, like this:

Parent component

```
<MyMenuBar>
    <button>Option 1</button>
    <button>Option 2</button>
</MyMenuBar>
```

MyMenuBar component

```
<template>
<div class="...">
    <slot></slot>
</div>
</template>
```

Provided that we applied the necessary styling and classes in MyMenuBar, the final rendered template may look something like this:

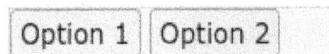

Figure 4.5 – A menu bar using slots

The logic that's applied is quite straightforward. The `<slot></slot>` placeholder will be replaced at runtime by whatever content is provided by the parent component inside the child tags. In the preceding example, if we inspect the rendered final HTML, we may find something like this (considering that we are using `W3.css` classes):

```
<div class="w3-bar w3-border w3-light-grey">
   <button>Option 1</button>
   <button>Option 2</button>
</div>
```

This is a fundamental concept in user interface design. Now, what if we need multiple "slots" – for example, to create a layout component? Here, an alternative syntax called *named slots* comes into play. Consider the following example:

MyLayout component

```
<div class="layout-wrapper">
    <section><slot name="sidebar"></slot></section>
    <header><slot name="header"></slot></header>
    <main><slot name="content"></slot></main>
</div>
```

As you can see, we have named each slot through the *name attribute*. In the parent component, we must now use the `template` element with the `v-slot` directive to access each one. Here is how a parent component would make use of `MyLayout`:

Parent component

```
<MyLayout>
    <template v-slot="sidebar"> ... </template>
    <template v-slot="header"> ... </template>
    <template v-slot="content"> ... </template>
</MyLayout>
```

The `v-slot` directive receives one argument, matching the slot name, with these remarks:

- If the name does not match any available slot, the content is not rendered.
- If no name is provided, or the name `default` is used, then the content is rendered in the default nameless slot.
- If no content is provided for a template, then the default elements inside of the slot definition will be shown. Default content is placed in between the slot tags: `<slot>...default content here...</slot>`.

There is also a shorthand notation for the `v-slot` directive. We just prefix the name of the slot with a numeral sign (#). For example, the templates in the preceding parent component can be simplified like so:

```
<template #sidebar> ... </template>
<template #header> ... </template>
<template #content> ... </template>
```

Slots in Vue 3 are very powerful, to the point that they even admit a way to pass props to the parent if needed. The syntax varies depending on whether we are using a *default slot* or *named slots*. For example, consider the following component template definition:

PassingPropsUpward component

```
<div>
    <slot :data="some_text"></data>
</div>
```

Here, the slot is passing a prop to the parent, named `data`. The parent component can then access it with the following syntax:

Parent component receiving props from the slot

```
<PassingPropsUpward v-slot="upwardProp">
    {{upwardProp.data}} //Renders the content of some_text
</PassingPropsUpward>
```

In the parent component, we use the `v-slot` directive and assign a local name to the props passed by the slot – in this case, `upwardProp`. This variable will receive an object similar in function to the props object but scoped to the element. Because of this, these types of slots are called *named scoped slots*, and the syntax is similar. Take a look at this example:

```
<template #header="upwardProp">
    {{upwardProp.data}}
</template>
```

There are other advanced uses for slots that cover edge cases, but we will not cover those in this book. Instead, I encourage you to investigate the topic further in the official documentation at `https://vuejs.org/guide/components/slots.html`.

There is one more concept related to this topic that we will see later in this book, in *Chapter 7, Data Flow Management*, that applies to reactive central state management. Now, let's look at some special components that behave a bit out of the ordinary.

Composables and mixins

In Vue 2, a special component named **mixin** allowed us to share code between components, thus avoiding code repetition. This approach created several issues and troubling side effects, whose solution evolved into creating the Composition API in Vue 3. The use of mixins is still supported for backward compatibility, but strongly discouraged. We will not cover mixins in this book; instead, we will focus on the technology that has replaced and surpassed them: `composables`.

A **composable** is a function that uses the Composition API to encapsulate and reuse *stateful logic* between components. It is important to distinguish composables from service classes or other encapsulations of *business logic*. The main purpose of a composable is to share *user interface or user interaction logic*. In general, each composable does the following:

- Exposes a function that returns *reactive* variables.

- Follows a naming convention prefixed with use in *camelCase* format – for example, useStore(), useAdmin(), useWindowsEvents(), and so on.

- It is self-contained in its own module.

- It handles *stateful logic*. This means that it manages data that persists and changes over time.

The classical example of a composable attaches itself to environmental events (window resizing, mouse movements, sensors, animations, and so on). Let's implement a simple composable that reads the vertical scroll of the document:

DocumentScroll.js

```
import {ref, onMounted, onUnmounted} from "vue"          //1
function useDocumentScroll(){
    const y=ref(window.scrollY)                           //2
    function update(){y.value=window.scrollY}
    onMounted(()=>{
        document.addEventListener('scroll', update)})     //3
    onUnmounted (()=>{
        document.removeEventListener('scroll', update)})  //4
    return {y}                                            //5
}
export {useDocumentScroll};                               //6
```

In this small composable, we start by importing component's life cycle events and the reactive constructor from Vue (//1). Our main function, useDocumentScroll, contains the entire code that we will share and export later (//6). In //2, we create a reactive constant and initialize it to the current window vertical scroll. Then, we create an internal function, update, that updates the value of y. We add this function as a listener to the document scroll event in //3, and then remove it in //4 (principle *"Clean after yourself,"* from *Chapter 2, Software Design Principles and Patterns*). Finally, in //5, we return our reactive constant wrapped in an object. Then, in a component, we use this composable in this way:

SomeComponent.js – script setup

```
import {useDocumentScroll} from "./DocumentScroll.js"
const {y}=useDocumentScroll()
...
```

Once we have imported the reactive variable, we can use it in our code and template as usual. If we need to use this bit of logic in more than one component, we just need to import the composable (**DRY** principle).

Finally, https://vueuse.org/ has an impressive collection of composables for our projects. It is worth checking out.

Dynamic components with "component:is"

The Vue 3 framework provides a special component called <component> whose job is to be a placeholder to render other components dynamically. It works with a special attribute, :is, that can receive either a **String** with the name of a component, or a variable with the component definition. It also accepts some basic expressions (a line of code that resolves to a value). Here is a simple example using an expression:

CoinFlip component

```
<script setup>
    import Heads from "./heads.vue"
    import Tails from "./tails.vue"
    function flipCoin(){return Math.random() > 0.5}
</script>
<template>
    <component :is="flipCoin()?Heads:Tails"></component>
</template>
```

When this component is rendered, we will see either the Heads or Tails component based on the result of the flipCoin() function.

At this point, you might be wondering, why not use a simple `v-show`/`v-if`? The power of this component becomes apparent when managing components dynamically and we don't know which ones are available at the time of creating the template. The official Vue Router, which we will see in *Chapter 5, Single-Page Applications*, uses this special component to simulate page navigation.

There is an edge case, however, that we need to be aware of. While most template attributes will pass through to the dynamic component, the use of some directives such as `v-model` will not work on *native input elements*. This situation is so rare that we will not discuss it in detail, but it can be found in the official documentation at `https://vuejs.org/api/built-in-special-elements.html#component`.

Now that we have a deeper understanding of components, let's put this new knowledge to work in two projects: a plugin, and a new version of our To-Do application.

A real-world example – a modals plugin

We have seen multiple approaches for sharing data and functionality *within* a project. Plugins are a design pattern for sharing functionality between projects and at the same time augmenting a system's capabilities. Vue 3 provides a very simple interface to create plugins and attach them to our application instance. Any object that exposes an `install()` method or a function that accepts the same parameters can become a plugin. A plugin can do the following:

- Register global components and directives
- Register an injectable resource at the application level
- Create and attach new properties or methods to the application

In this section, we will create a plugin that implements modal dialogs as global components. We will use dependency injection to provide them as a resource and leverage Vue's reactivity to manage them through promises.

Setting up our development project

Follow the instructions in *Chapter 3, Setting Up a Working Project*, so that you have a starting point. In the `src/` directory, create a new folder named `plugins/`, with a sub-folder named `modals/`. It is a standard approach to place our plugins in individual directories inside the `plugins/` folder.

The design

Our plugin will install a component globally and keep an internal reactive state to track the current modal dialog status. It will also provide an API to be injected as a dependency to those components that need to open a modal dialog. This interaction can be represented like this:

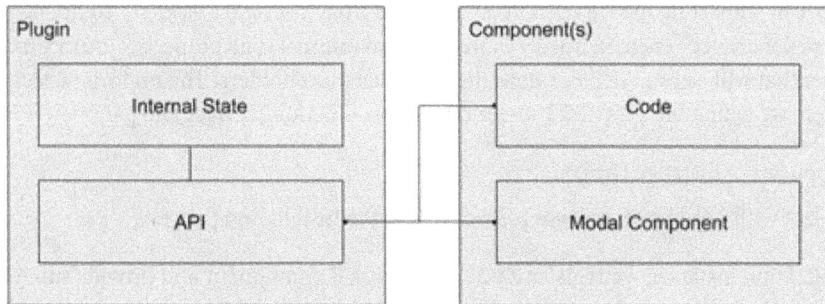

Figure 4.6 – The modal plugin's representation

Components will implement a modal element, and we will open the dialog through code. When a modal is open, it will return a promise following the async pattern. When the user accepts the modal, the promise will resolve, and reject upon cancellation. The content of the modal will be provided by the parent component through the use of slots.

The implementation

For this plugin, we will only need two files – one for the plugin's logic and one for our component. Go ahead and create the index.js and Modal.vue files in the src/plugins/modal folder. At the moment, just scaffold the component with the section's script setup, template, and style. We will come back later to complete it. With those files in place, let's start, step by step, with the index.js file:

/src/plugins/modals/index.js

```
import { reactive } from "vue"                          //1
import Modal from "./Modal.vue"
const
    _current = reactive({}),                            //2
    api = {},                                           //3
    plugin = {
        install(App, options) {                         //4
            App.component("Modal", Modal)
            App.provide("$modals", api)
        }
    }
export default plugin
```

We start in //1 by importing the reactive constructor from Vue, and a Modal component whose file we have not yet created. Then, in line //2, we create an internal state property, _current, and in //3, an object that will be our API. For now, these are just placeholders. The important section is in line //4, where we define the install() function. This function receives two parameters in order:

1. The application instance (App).

2. An object with options, if one was passed during the installation process.

With the application instance, we register Modal as a global component and provide the API as an injectable resource under the name $modals, both at the application level. To use the plugin in our application, we must import it into main.js and register it with the use method. The code looks like this:

/src/Main.js

```
import { createApp } from 'vue'
import App from './App.vue'
import Modals from "./plugins/modals"
createApp(App).use(Modals).mount('#app')
```

As you can see, creating and using a plugin is rather simple. However, thus far, our plugin doesn't do much. Let's go back to our plugin code and complete the API. What we need is the following:

- A show() method that takes a name that identifies a modal dialog implementation and returns a promise. We will then save the name and references to the resolve() and reject() functions in our reactive state.

- An accept() and cancel() methods, to resolve and reject the promise, respectively.

- An active() method to retrieve the name of the current modal.

Following these guidelines, we can complete the code so that our index.js file looks like this:

/src/plugins/modals/index.js

```
import { reactive } from "vue"
import Modal from "./Modal.vue"
const
_current = reactive({name:"",resolve:null,reject:null}),
api = {
        active() {return _current.name;},
        show(name) {
                _current.name = name;
                return new Promise(
```

```
            (resolve = () => { }, reject = () => { }) => {
                _current.resolve = resolve;
                _current.reject = reject;
            })
        },
        accept() {_current.resolve();_current.name = "" },
        cancel() {_current.reject();_current.name = "" }
    },
plugin = {...} // Omitted for brevity
export default plugin;
```

Our internal state is kept with a `reactive` variable and only accessed through our API. In general, this is a good design for any API. Now, it is time to make the magic happen in our `Modal.vue` component, to complete the workflow. I'm omitting the classes and styles for brevity, but the full code can be found in this book's GitHub repository at `https://github.com/PacktPublishing/Vue.js-3-Design-Patterns-and-Best-Practices`.

Our modal component will have to do the following:

- Cover the entire viewable area with a translucent element to block interaction with the rest of the application

- Provide the dialog to be displayed:

  - A *prop* to register the name of the component, as provided by the parent.

  - A *header* to display a title. The title will also be a prop.

  - An area for the parent component to populate with custom content.

  - A footer with *accept* and *cancel* buttons.

  - A reactive property that triggers when the component should appear.

With our definition in place, let's work on the template:

/src/plugins/modals/Modal.vue

```
<template>
<div class="viewport-wrapper" v-if="_show">                  //1
  <div class="dialog-wrapper">
   <header>{{$props.title}}</header>                         //2
   <main><slot></slot></main>                                //3
   <footer>
      <button @click="closeModal(true)">Accept</button>      //4
      <button @click="closeModal(false)">Cancel</button>
   </footer>
```

```
    </div>
  </div>
</template>
```

In line //1, the reactive variable, _show, controls the visibility of the dialog modal. We display the prop title in line //2, and reserve a slot in line //3. The buttons in line //4 will close the modal on the click event, each one with a representative Boolean value.

Not, it's time to write the logic of the component. In our script, we need to do the following:

- Define two props: title (for display) and name (for identification)

- Inject the $modals resource so that we can interact with the API and do these things:

  - Check if the modal's name matches the current component (this "opens" the modal dialog)

  - Close the modal by resolving or rejecting the promise

Following these directions, we can complete our script setup:

```
<script setup>
  import { inject, computed } from "vue"                   //1
  const
  $props = defineProps({                                    //2
      name: { type: String, default: "" },
      title: { type: String, default: "Modal dialog" }
      }),
  $modals = inject("$modals"),                              //3
  _show = computed(() => {                                  //4
      return $modals.active() == $props.name
  })
  function closeModal(accept = false) {
      accept?$modals.accept():$modals.cancel()             //5
  }
</script>
```

We begin in line //1 by importing the inject and computed functions. In line //2, we create the props with sensible defaults. In line //3, we inject the $modals resource (dependency) that we will use in the computed property in line //4 to retrieve the current active modal and compare it with the component. Finally, in line //5, based on the click of the buttons, we trigger the resolution or rejection of the promise.

To use this plugin from any component in our application, we must follow these steps:

- In `template`, define a modal component with the name registered in our plugin (`Modal`). Notice the use of the attributes for props:

```
<Modal name="myModal" title="Modal example">
      Some important content here
</Modal>
```

- In our script setup, inject the dependency with the following code:

```
const $modals = inject("$modals")
```

- Display the modal component by the given name with this code:

```
$modals.show("myModal").then(() => {
      // Modal accepted.
}, () => {
      // Modal cancelled
})
```

With this, we have completed our first plugin in Vue 3. Let's put it to good use in our new To-Do list application.

Implementing our new To-Do application

At the beginning of this chapter, we saw a design for our new to-do application, and we sliced it into hierarchical components (see *Figure 4.1*). To follow the rest of this section, you will need a copy of the source code from this book's GitHub repository (`https://github.com/PacktPublishing/ Vue.js-3-Design-Patterns-and-Best-Practices`). As our code base grows, it is not possible to see each piece of implementation in detail, so we will focus on the main changes and specific bits of code. With that in mind, let's review the changes from the previous implementation, roughly in order of file execution. To start, we added two new directories to our project:

- `/src/plugins`, where we placed our `Modals` plugin.
- `/src/services`, where we place modules with our business or middleware logic. Here, we created a service object to handle the business logic of our To-Do list: the `todo.js` file.

In `main.js`, we import and add our plugin to the application object, using the `.use(Modals)` method to register our plugin.

The `App.vue` file has become primarily a layout component, without any other application logic. We import and use a header, (`MainHeader.vue`), and a parent component to manage our To-Do list and UI, (`ToDoProject.vue`), just like in the design shown in *Figure 4.2*.

The `ToDoProject` component contains the state of the list through reactive variables, where we have the following:

- `_items` is an array that contains our To-Do items

- `_item` is an auxiliary reactive variable that we use to create new items or edit a duplicate of an item

- `_filter` is another auxiliary reactive variable that we use to input a string to filter on our list

It is worth mentioning that we also declare a constant, `$modals`, that receives the injected `Modals` object API. Notice how the `showModal()` function opens and manages the result of the dialog for new and edited items using this object. Then, the modal in question appears in the template, by the ending marked with a comment. It is customary to place all the modal templates toward the end of a component, instead of being spaced all over the template.

The `ToDoProject` component delegates state data through props to child components to display summary and list items. It also receives events from them, with instructions to manipulate the list. You can consider this component as the *root* of functionality. Our application only has one, but this begins to hint at the idea of how a web application starts to get organized by functionality.

There is another point worth mentioning, which is the use of *service objects and classes*. In our application, we have `todo.js`, which we import as `todoService` where needed. In this case, this is a *singleton*, but it could also be a class constructor. *Notice that it does not contain any interface logic, only application or business logic*. This is a defining factor that differentiates it from *composables*, which we have seen before.

Another change is that the To-Do items now have multiple states, and we can cycle through them with a click. We have implemented this logic in the `toggleStatus()` function of the service, *not in the component*. The flow between the states can be represented like this:

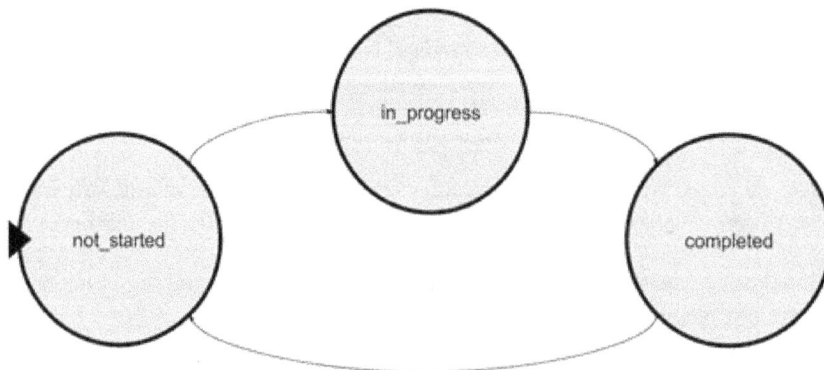

Figure 4.7 – A circular finite state machine

You may recognize the design, as it represents a **circular finite state machine**. *Finite state machines are very handy for representing the possible states of an item, and the conditions that trigger each change* (in our case, a user's click). There are many ways to implement a state machine, but one of the simplest is with a `switch` statement, like in our example:

Todo.js

```
[function] toggleStatus(status){
    switch(status){
        case "not_started":     return "in_progress"
        case "in_progress":     return "completed"
        case "completed":       return "not_started"
    }
}
```

This function, given its current status, will return the next one. Calling this function on each click, we can update the state of each item in a clean manner.

The final point to remark about this new implementation is the use of computed properties in the ToDoSummary component. We use them to display summary cards with the different states of our items. Notice how well the reactivity works – the moment we alter the state of an item in the list, the summary is immediately updated!

With the new implementation in order, it is time to take a step back and view our work with a critical mind.

A small critique of our new To-Do application

This new version of the To-Do application is a clear improvement over our first approach, but it can be improved:

- We still have only one list of tasks.
- Everything still happens in just one page.
- Our items are ephemeral. They disappear when we close or refresh the browser.
- There is no security, no way to have multiple users, and so on.
- We can only add plain text. How about images or rich text?
- With a bit of work, we could expand our application so that it manages multiple projects, additional content, assignments, and more.
- We have made good progress, but there is still more to do.

Summary

In this chapter, we dived deep into components and learned how they can communicate, share functionality, and implement design patterns within the framework. We also saw an approach to transform a rough sketch or detailed design into components. We then learned about special components, created a plugin for modal dialogs using the framework's dependency injection, and applied other patterns to make our coding easier and more congruent. Furthermore, we refactored our application and expanded its capabilities while taking a glimpse at better state management, independent from the HTML element we used before. We have made good progress, but there is still more to go.

In the next chapter, we will create a **single-page application** (**SPA**) with what we learned this far.

Review questions

Answer the following questions to test your knowledge of this chapter:

- How can we start from a visual design or prototype and plan the implementation with components?

- What are the many ways that components can communicate with each other?

- How we can reuse code in multiple components? Is there any other way?

- What is a plugin, and how can we create one?

- Which patterns have we applied to the new To-Do application?

- What would you change in the implementation?

5

Single-Page Applications

In this chapter, we continue to increase our skills in Vue 3 with the introduction of **single-page applications** (**SPAs**). We will learn what distinguishes them from regular websites and will dive into their key characteristics. To put this into action, we will build a new version of our To-Do application using the Vue Router and a different pattern of communication than the previous chapters. We will also learn authentication methods with code samples.

By the end of this chapter, you will know the following:

- How to create SPAs with Vue 3

- How to organize your application to make use of the Vue Router with different routing strategies

- How to reimplement our To-Do application with a practical application of different patterns

- How to implement different patterns of authentication in your SPA

While the previous chapter was somewhat heavy with foundational knowledge, from now onward, we center more on practical matters. Because of this, you will need access to the example applications to follow through.

Technical requirements

The code for this chapter can be found on GitHub, at `https://github.com/PacktPublishing/Vue.js-3-Design-Patterns-and-Best-Practices/tree/main/Chapter05`.

Check out the following video to see the Code in Action: `https://packt.link/RnAyz`

What is a SPA?

To explain what a SPA is, we should first explain how we interact with the **World Wide Web** (**WWW** or **W3**). When we enter an address in a web browser, we receive a web page sent by a web server. In the most basic form, a website is just a collection of pages, mostly what we call "static pages". Static in this context means that the same files in the server are sent without modification. This makes a website very fast and secure. However, a purely static site does not offer much interactivity with the end user. Sometimes this is referred to as **Web 1.0**. Server and browser scripting came in to solve this limitation and gave birth to **multi-page applications** (**MPAs**). Pages could now be either static or dynamically generated on the server, which in turn could also receive calls for new pages with additional data that processes them and returns a new page in response. These new pages "generated on the fly" are called **dynamic** and made it possible to have applications. These technologies made it possible for blogs, services, and businesses to proliferate.

It was with the introduction of key technologies such as asynchronous communications (**AJAX**), more powerful JavaScript, local storage methodologies, increased network speeds, and computational power that we came to what is known as **Web 2.0**. It was now possible to load a single file into the browser and use JavaScript to take control of the entire interface and interactivity, producing rich and heavily interactive applications without generating new pages on the server. The SPA only contacts the server to load bits of data, the UI, and so on, as needed. It is now possible to migrate to web technologies what were traditional "desktop-only" applications, such as text editors, spreadsheets, rich email clients, graphic design suites, and so on. *Office 365*, *Google Docs*, *Photoshop online*, *Telegram*, *Discord*, *Netflix*, *YouTube*, and so on are good examples of SPAs. It is important to acknowledge that the introduction of SPAs does not invalidate the use of MPAs or make them obsolete—each has its utility in certain contexts. Most blogs and news sites today are, in fact, MPAs and still constitute a significant part of the internet. The most complex web applications today include a mixture of MPAs and SPAs, working together. SPAs can even be installed as hybrid applications on desktop and mobile devices. We will see how to implement this in *Chapter 6, Progressive Web Applications*.

Moving forward, with the explosion of distributed and decentralized computing, and smart blockchains, the technology that makes up SPAs has gained even more relevance. While not fully rooted in general use, this new era in web evolution is called **Web 3.0**. We will see in this chapter more about this topic, with examples.

All the applications that we have made this far fall into the SPA category, even if we have not used their full potential yet. Vue 3 is specially designed to create these types of applications, and is one of the most relevant technologies for such an approach, together with *React*, *Angular*, *Svelte*, and others. But not everything is sugar, glitter, and rainbows. As with any technology, there are trade-offs to using SPAs. In the next table, we list some of them:

Advantages	Disadvantages
<ul><li>Faster and smarter loading times</li><li>Local caching for improved performance</li><li>Rich UIs and interactivity</li><li>Easier to develop and test than MPAs</li><li>More efficient use of code and templates, with less network communication (as compared to full-page re/loads)</li></ul>	<ul><li>Difficult for search engines to index or discover</li><li>Increased complexity</li><li>Increased load time and slower time for first interactivity</li></ul>

Table 5.1 – Advantages and trade-offs for SPAs

As you can see, the list of advantages is by far more important than the disadvantages. You should consider using a SPA when the application requires significant user interactivity and real-time feedback. Now that we have a better idea of what a SPA is, let's see the key concept at the core of their functionality: the application **router**.

The Vue 3 router

Vue is a great framework to build SPAs, but without a router, the task would soon become quite complex. The Vue router is an official plugin that takes over the navigation of the application and matches a URL to a component. This gives us the advantages of an MPA. With the router, we can do the following:

- Create and manage dynamic routes to components, matching parameters to props automatically if needed

- Identify routes (addresses and components) by name and trigger navigation by code

- Load components dynamically when needed, thus reducing the bundle size

- Create a natural and logical way to approach website navigation and code splitting

- Control navigation with well-known events, before and after navigation occurred

- Create page transition animations in a way that is not possible with MPAs

The implementation of the Vue 3 router is simple and follows the same methodology as with other components of the ecosystem. Let's take our project from *Chapter 4, User Interface Composition with Components*, and modify it to use the Vue router.

Installation

When starting a new project, you may have noticed that the installer menu gives you the option to install the Vue router. If you have not selected this option, as we did in our example app, the installation afterward is quite simple. In a terminal, in the project directory, just execute the following command:

```
$ npm install vue-router@4
```

The command will download and install the dependencies, just as with any other package in the `node_modules` directory. In order to use it in our application, we need to do the following:

1. Create our routes.
2. Link the routes to our components.
3. Include the router in our application.
4. Set our templates where the router will display our components.

As with much of the framework, the router does not specify in which directories or organization your routes should be placed, or your components for that matter. However, there is a convention that we will use that has become the de facto standard in the industry. In the `/src` folder, create the following directories:

- `/router` (or `/routes`): Here, we will have our JavaScript files with the routes for our application
- `/views`: This folder will contain the top-level components that match the application navigation (as a best practice)

With these directories in place, we are ready to start modifying our application to include route navigation. Before that, let's take a look at what we want to achieve with our router.

A new To-Do application

Our new application will reuse the components to display our To-Do list, but also will accommodate the creation of multiple lists or projects. We will display a sidebar with all our projects, and when selecting them, the list will be updated11. These projects will also be persisted in the browser, so we can come back to them later by using `localStorage`. We will then have a very simple navigation, with two top-level pages (components):

- A landing page where we can create new projects
- A project page where we can work with our to-do list

Following these simple premises, our application once finished will look like this:

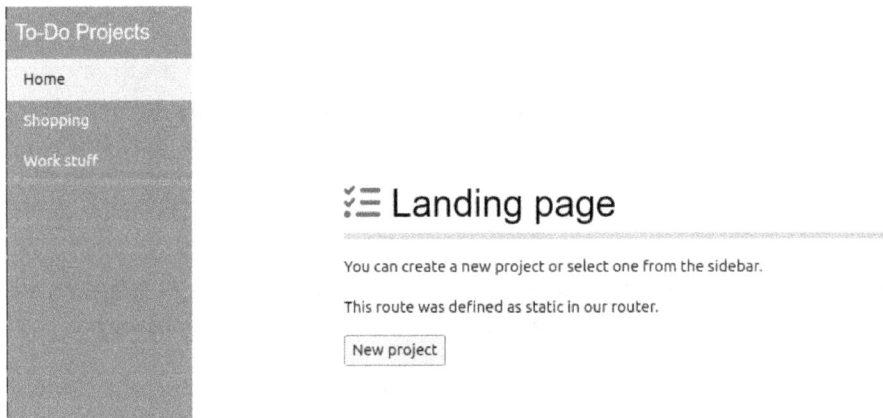

Figure 5.1 – Our landing page

As you can see in *Figure 5.1*, the landing page is also the place where we can create new projects. We use modal dialogs to collect user input, just as we did before. On the sidebar, we display a link to the **Home** page (the landing page) and a list with all the names of the different projects that we have created. When you click on each one, the route in the browser (URL) will update as well as the page, and we will see something like this:

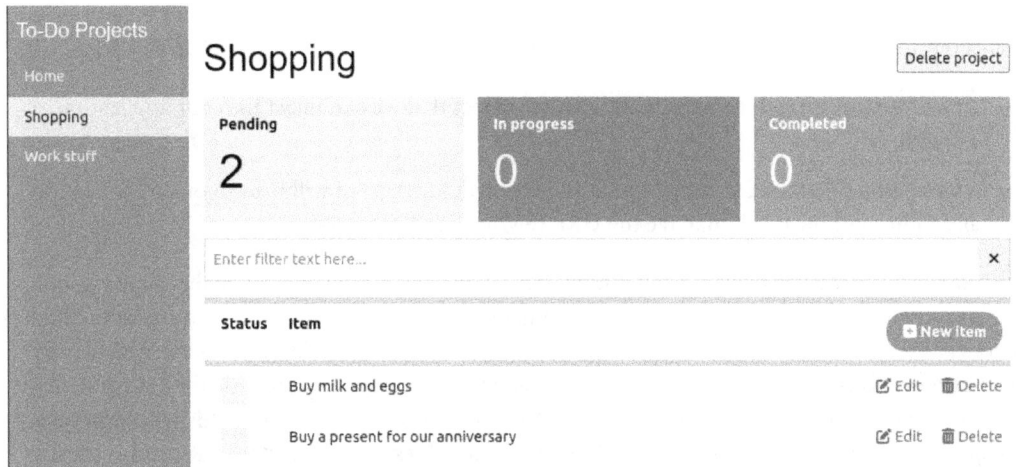

Figure 5.2 – A To-Do project page

You may recognize this last screenshot, as it is what our `ToDoProject.vue` component displays. As a matter of fact, it will require very little modification to reach this result. For now, let's begin with the routes.

Routes' definition and the Router object

To create routes for our project, we need to first define them in their own module. In the `/router` directory, create an `index.js` file with the following content:

/src/router/index.js

```
import {createRouter,createWebHashHistory} from 'vue-router'    //1
import Landing from "../views/Landing.vue"                      //2
const routes = [
    {path: "/",name: "landing",component: Landing},
    {path: "/project/:id",name: "project",
        component: () =>
            import("../views/ToDoProject.vue"),                //3
        props: true
}],
    router = createRouter({                                     //4
        history: createWebHashHistory(),                        //5
```

```
    routes,
    scrollBehavior(to, from, savedPosition){return{top:0}}
})
export default router;
```

We start our file by importing two constructors from the `vue-router` package, in line `//1`:

- `createRouter`, which will create a router object that we can inject into our application as a plugin
- `createWebHashHistory`, which is a constructor that we will pass to our router object and indicates how it will manage the URL rewriting in the browser

`Web hash history` will display # (a numeral sign) in the URL and will indicate that all navigation points to a single file. All navigation and URL parameters will follow this sign. It is the easiest method and does not require any special configuration. However, the other available methods are **Web history** (also known as *HTML5 mode* or *pretty URLs*) and **Memory**. Web history does not use the hash notation, but does require a special server configuration. We will see how to accomplish this with examples in *Chapter 10, Deploying Your Application*. Memory mode does not modify the URL and is mostly used for web views (as in hybrid frameworks such as NW.js, Electron, Tauri, Cordova, Capacitor, and so on) and **server-side rendering (SSR)**. For now, we will stay with the **Web hash history** method.

In line `//2`, we import a component using the static notation, and we define a `routes` array with our routes. Each route is represented by an object with at least the following fields:

- `path`: A string that represents the URL associated with the component
- `name`: A string that behaves like a unique ID for the route and that we can call programmatically
- `component`: The component to render

Notice how in line `//2` we import a static component, but in line `//3`, we use the dynamic import notation. This implies that the first route (named `"landing"`) will be included in the main bundle, but the second route (in line `//3`, named `"project"`) will only be loaded the first time it is needed, from a separate bundle. Using routes, we can create a strategy for improving our application loading and bundle size.

Finally, in line `//4`, we create our `router` object using the constructor and passing an options object. Notice in line `//5` how we pass the `history` field a constructor for our chosen `history` method. We also pass our routes (obviously), and also as an example here, we create one of the possible *navigation guards*, to make sure that after navigating to each route, the window scrolls all the way to the top. Without this, we may encounter a strange side effect, with the scroll not changing between "*pages*." Navigation guards are triggered before and after a navigation event. They can be used in a multitude of situations, such as authentication control or data preloading. Please refer to the official documentation for a complete list of guards, with examples (`https://router.vuejs.org/guide/advanced/navigation-guards.html`).

In our second route, we have also included a variant in the notation of the path, with the inclusion of a named parameter prefixed by a semicolon (:id). This route will match anything following /project/ and assign it to a reactive variable, which we can access programmatically (we will see how this works later). The route also has an additional field, props: true. This indicates that the parameter named in the path will be automatically passed as a prop to the component if the component has defined a prop with the same name. This will become useful and apparent in the next sections.

With our routes and router defined, it is time to import them into our main.js file and attach them to our application. The file will look now like this:

/src/main.js

```
import { createApp } from 'vue'
import router from "./router"
import App from './App.vue'
import Modals from "./plugins/modals"
import styles from "./assets/styles.css"
createApp(App).use(router).use(Modals).mount('#app')
```

Easy enough—it is now time to create components that are currently missing and adapt the ones we have. Before dealing with the code, let's see which new components the router provides our application with.

Router template components

When we include the router in the application, it injects into the global scope the following new components:

- RouterView: This component provides the placeholder where the route components will be rendered.

- RouterLink: Provides an easy way to link to routes; through the use of handy props and styles, we can control the appearance and final render element.

Together with the router and routes definition, these two components in our template make it possible to offer navigation and better organize our code. Before we dig into their details, let's see them in action in our application. Let's start modifying our App.vue component to turn it into a layout container (styles omitted):

App.vue

```
<script setup>
    import Sidebar from './components/Sidebar/Sidebar.vue';
</script>
```

```
<template>
<div class="app">
   <Sidebar></Sidebar>
   <main>
      <router-view></router-view>
   </main>
</div>
</template>
```

As you can see, we include a new component, `Sidebar`, which will contain the main navigation for our application. Then, we just place a single `<router-view>` component, where our router will render each page. When it comes to the styles, I will refer to the code in GitHub for the details. Now, it is time to create the `Sidebar` component in the `/src/components/Sidebar/Sidebar.vue` path and copy the code from the repository. There is a lot to see in this small file. Let's start looking into the template and how we use the `RouterLink` instances. The first one is static and points to the landing page. Instead of just using a link or an anchor tag, we define the target of the link as an object where we reference the name of the route directly:

```
<RouterLink :to="{name:'landing'}" class="w3-padding" active-
class="w3-yellow">Home</RouterLink>
```

When this component is rendered, by default, it will become an anchor tag, and the `href` attribute will be dynamically converted to the appropriate route. If we change our route's definition and give it another path, it won't affect this code. It is a good practice to reference routes by their names, instead of by their URL. In the case that we need to pass some query string parameters to the URL, we can easily do it by passing an object with key/value members as a `params` attribute. Here is an example:

```
<RouterLink :to="{name:'search',params:{text:'abc' }}" >Search</
RouterLink>
```

The preceding `params` attribute will be rendered as a URI with the `?text=abc` query string. As we mentioned, if the route has the `props` attribute active and the receiving component has defined a prop of the same name, the value will be automatically assigned. This is a situation that allows us to generate a list of links and pass to our project page the ID of each project, as you can see next in the file:

```
<div v-for="p in _projects" :key="p.id">
   <RouterLink :to="{name:'project',params:{id:p.id}}">
         {{p.name}}
   </RouterLink>
</div>
```

When we create a project on the landing page, we automatically assign a unique ID to each one, which we use in the previous code. Just as with other props, we can watch the changes and react by loading the respective To-Do items for each project. With that in mind, we modified the `ToDoProject.vue` file to define the prop (no need to define the type):

```
$props=defineProps(["id"])
```

And then, we also set a watcher to detect changes with these lines in the `script` section:

```
import { watch } from "vue"
watch(()=>$props.id, loadProject)
```

This watch receives a function that returns the `prop` attribute and then runs the `loadProject()` function. At this point, you may ask why we need to do this since each URL is different. The answer is that Vue and the router only load a component the first time it's needed. As long as it remains in view, it doesn't reload it and only updates reactive properties. Since our `script setup` code only runs during the first load, at the moment of creation, we need a way to detect changes to run non-reactive operations, such as loading the To-Do items for the project from `localStorage`.

You can follow the rest of the changes in the repository. There is very little that changes in the components that work with the To-Do list, and that is the point of the encapsulation. Even the modification of `ToDoProject.vue` is small. However, there is one design decision that we need to point out: the use of the *pub/sub model* to keep the sidebar menu synchronized.

We have created a singleton with an event bus (`eventBus`). When we create a new project or delete it, we trigger an update event with this line:

```
eventBus.emit("#UpdateProjects")
```

We register the listening events in those components that need it during the *mounting* lifecycle event of the component, and we de-register it before is *unmounted*. In our case, we only need this in the `Sidebar` component, but we could have it anywhere in our application as needed:

```
onMounted(()=>{
    eventBus.on("#UpdateProjects", updateProjects)
})
onBeforeUnmount(()=>{
    eventBus.off("#UpdateProjects", updateProjects)
})
```

The name of the event is trivial and does not follow any convention. In this book, we prefix it with a numeral sign, as a personal preference.

In previous implementations, as well as in the `ToDoProject.vue` component, we use the parent as the conduit to share information between sibling components, as we discussed previously. Here, we use another model, the *pub/sub pattern*, to avoid polluting the `App.vue` component with such a task. In *Chapter 7, Data Flow Management*, we will see other approaches for central state management. Let's now take a closer look into more examples and details of using the router with more advanced scenarios.

Nested routes, named views, and programmatic navigation

This far, we have created static and dynamic routes, even with some parameters in the address. But the router can do even more than that. By using named routes, we can also create "sub-routes" and named "sub-views" to create deeper navigation trees and complex layouts.

Let's start with an example. Suppose we have a data structure on three levels, and we want to reactively present this to the user in such a way that they can choose one level, and then "drill down" to the details. We also want to have this reflected in the URL, in such a way that we can share or reference the full case. The levels, in this case, would be country, state, and city. The UI would then look something like this:

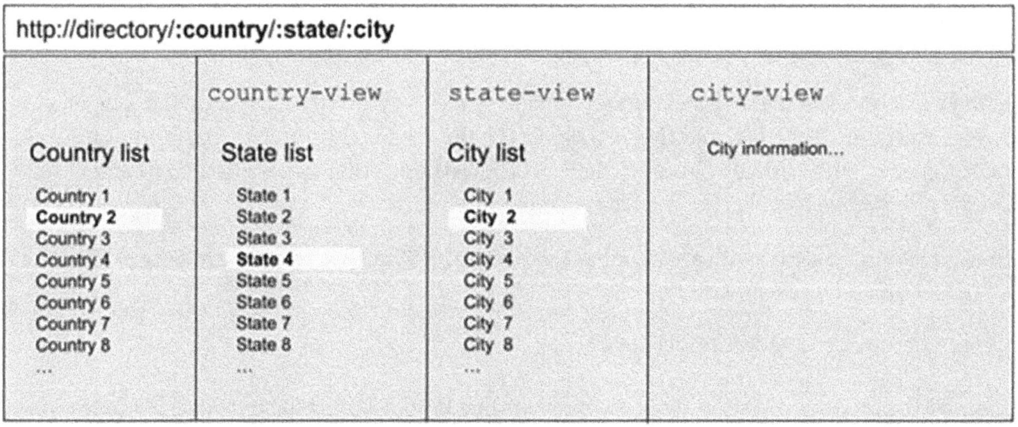

Figure 5.3 – A selection using multiple named views and sub-routes

As you can already guess from the screenshot, when the user selects the country, the state list is populated, and the URL is updated. When selecting a state, the city list is updated... and finally, when selecting the city, the information appears in the last column. You may have seen this method of navigation before. There are multiple ways to implement this, some more efficient than others. Our intent is to implement this as a learning exercise, so let's start with the routes' definition. Here is a segment of our routes' definition array:

```
{
path: "/directory", name: "directory",
component: () => import("../views/Directory.vue"),
children:[
{ path:":country", name: "states", props: true,
  component: ()=>import("../views/State.vue"),
  children:[
      { path:":state", name: "cities", props: true,
        component: ()=>import("../views/City.vue")
```

```
            } ]
} ] }
```

Nested routes' definition

At first sight, you will notice that not much has changed, save for the inclusion of a new attribute on the route: `children[]`. This attribute receives an array of routes, which in turn can have other children, as we see in the previous code snippet. Children routes will be rendered in the `RouteView` component of their parents, and their paths will be concatenated with their parents as well, unless they start with the root (with a backslash).

To navigate to each route, we could use any of the methods recognized by the router. However, it is a good practice to use their names and pass any parameter or query string through an object, and let the router resolve the URL. As an example, see how in the `Directory.vue` component we use the `RouterLink` element:

/src/views/Directory.vue component, lines 13-18

```
<div v-for="c in countries" :key="c.code">
<RouterLink
    :to="{name:'states', params:{country:c.code}}"
    active-class="selected">
    {{c.name}}
</RouterLink>
</div>
```

We have included our `RouterLink` component inside of a loop, to create as many links as needed based on our data. The target of the link is set to an object, where we pass the name of the route (`states`), and pass parameters respecting the route and props definition for the component. Notice that the path of the component has been defined as a parameter (it starts with a colon character—`:country`) and it also matches the props definition of the object in `State.vue`. This correlation is what enables the router to automatically pass the data for us.

As you inspect the code, you will notice that in our smallest child component, the `City.vue` file, we define in our props both country and state. However, in the route definition, only one parameter appears: the state (`:state`). Nevertheless, when you run the example, you will notice that the prop is also populated. This happens because children components inherit, together with the URL path, all the parameters defined in the route of the parents. In this case, our component then also receives the `:country` parameter that was passed to the parent, even if it doesn't show up in its specific route.

When you run the application, you will see something similar to this screenshot:

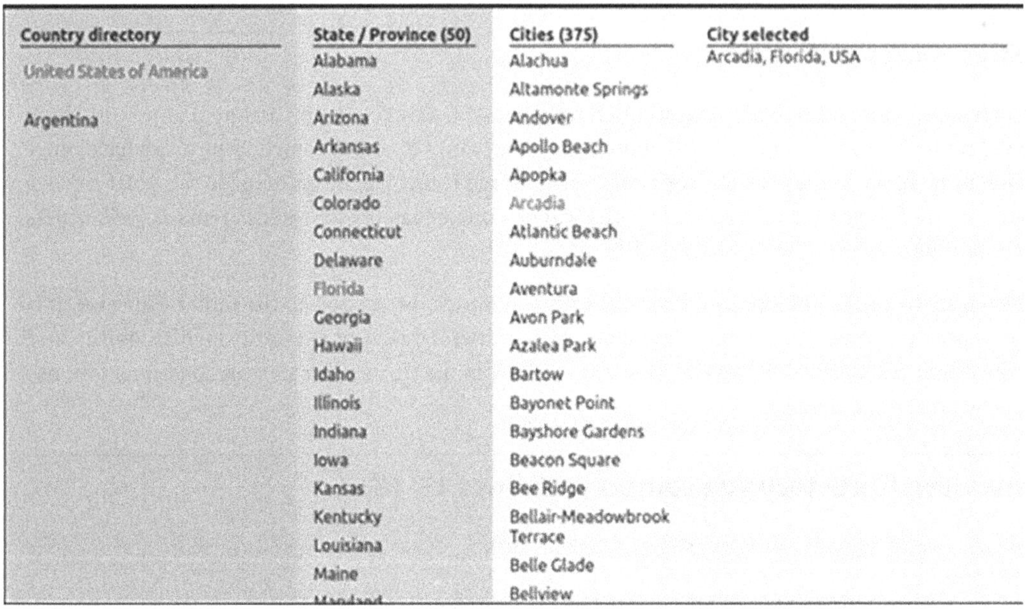

Country directory	State / Province (50)	Cities (375)	City selected
United States of America	Alabama	Alachua	Arcadia, Florida, USA
	Alaska	Altamonte Springs	
Argentina	Arizona	Andover	
	Arkansas	Apollo Beach	
	California	Apopka	
	Colorado	Arcadia	
	Connecticut	Atlantic Beach	
	Delaware	Auburndale	
	Florida	Aventura	
	Georgia	Avon Park	
	Hawaii	Azalea Park	
	Idaho	Bartow	
	Illinois	Bayonet Point	
	Indiana	Bayshore Gardens	
	Iowa	Beacon Square	
	Kansas	Bee Ridge	
	Kentucky	Bellair-Meadowbrook Terrace	
	Louisiana		
	Maine	Belle Glade	
	Maryland	Bellview	

Figure 5.4 – Nested routes example, with selections

Only two countries have been included from static files, for simplicity. In a real-life project, this data would be retrieved from a database.

We have used until now "default" `RouteView` components, but the Vue router allows us to include multiple views in one component, by assigning them different names. We will only see the notation here, as the implementation is trivial. Consider a component with the following template:

```
<div>
    <RouterView name="header"></RouterView>
    <RouterView name="sidebar"></RouterView>
    <RouterView></RouterView>
</div>
```

In the preceding code, we give our routes an identification with the `name` attribute. We also have a view without a name, in which case it is considered the "default" view, or with the name `default` as well. To make use of this new layout, the routes' definition changes slightly. In each definition now, we do not have a `component` attribute, but instead, a `components` (in plural) attribute that expects an object. Each field's name in the object must match the names given to our `RouterView` components and be equal to an object. For the previous piece of code, the equivalent route definition would be something like this:

```
{ path:"/layout", name: "main",
  components:{
      default: ()=>import('...'),
      header: ()=>import('...'),
      sidebar: ()=>import('...')
}}
```

Using this type of definition, we can create complex layouts, as we can also define sub-routes to make use of—for example—the header and sidebar from the parent and only render in the default view. We have an impressive number of possibilities for building dynamic UIs.

One important topic that we must cover before moving to the next section is that of programmatic navigation. We have used thus far the new components provided by the router, but we can also trigger navigation directly from our JavaScript without having to rely on the user triggering an event. For this, the Vue Router provides us with two handy constructors to use in our components' scripts: `useRoute` and `useRouter`. We import these constructors into our components with the following line:

```
import {useRoute, useRouter} from "vue-router"
const       $route=useRoute(),
            $router=useRouter()
```

As you can imagine, `$route` provides us with information about the current route, while `$router` allows us to modify and trigger navigation events.

The $router object provides several methods, of which the most often used are summarized in this table:

Method	Description

.push()	The most important method. It pushes a new URL into the web history and navigates to the corresponding component. It is the programmatic equivalent of using RouterLink. It accepts either a string with the URL to navigate or an object with optional attributes. Here are some examples for each accepted parameter:
	
	``// Navigate to an URL
	$router.push("/my/route")
	
	// Navigate to a URL, using an object
	$router.push({path: "/my/route"})
	
	// Navigate to a route, with parameters
	$router.push({
	    name:"route-name",
	    params:{key:value}
	})
	
	// Navigate to a route, with query strings
	$router.push({
	    name:"route-name",
	    query:{key:value}
	})```
	
	Of course, you can create complex routes by passing parameters and query strings. What is important to remember is that .push will update the navigation history in the browser.

.replace()	Replace the current navigation component, without modifying the URL. It accepts the same parameters as .push.

.go()	This method receives an integer number as a parameter and triggers navigation using the browser's history. Positive numbers navigate forward and negative numbers go backward in the navigation history. Its most common use is for implementing a "go back" link in an application. Here are some examples:
	
	``// Go back one entry
	$router.go(-1)
	
	// Go forward one entry
	$router.go(1)```

As mentioned, these are the most commonly used methods and the ones you should have present. I can say that using these will cover the vast majority of regular necessities. A full list of methods available can be found in the official documentation and allow you to manage also edge cases that may arise. I encourage you to check them out, at least to be aware of them, at `https://router.vuejs.org/api/interfaces/Router.html#properties`. Some of these edge cases could be: add and remove routes dynamically (`.addRoute()` and `.removeRoute()`), retrieve the registered routes (`.getRoutes()`), check whether a route exists before navigating to it (`.hasRoute()`), and so on. We will not use them, so it is not relevant to see them in detail here.

In contrast, the `$route` object gives us information about the current path (URL) where our component is being rendered. As with the previous example, here is a list of the most commonly used attributes, and their function:

Attributes	Description
`.name`	Returns the current name of the route.
`.params`	Returns an object with the parameters provided with the path (URL). If these have been matched to props, the values may overlap.
`.query`	Returns an object with the decoded query string attached to the current path.
`.hash`	If any, it returns the path in the URL following and including the hash sign (#).
`.fullPath`	Returns a string with the full path of the route.

In the examples of this book, we will use `.name()`, `.params()`, and `.query()` on more than one occasion, as they tend to be the most commonly used as well. A full list of methods and properties can be found in the official documentation.

> **Important notation differences**
>
> We have been using the `useRoute` and `useRouter` constructors in the Composition API with the `script setup` notation. In the Options API, there is no need to initialize these objects. Both are available automatically through `this.$route` and `this.$router`. Also, the `$route` and `$router` objects are available automatically in the template, when using the Composition API.

A full code example can be found in the GitHub repository, under `Chapter 5/Nested Routes`, at this URL: `https://github.com/PacktPublishing/Vue.js-3-Design-Patterns-and-Best-Practices/tree/main/chapter05`.

Now that we know how to handle routes, parameters, and query strings, it is time to look into some common patterns for authentication in SPAs, since different paths (URLs) are necessary for many of them.

Exploring authentication patterns

The power of SPAs becomes apparent when there is also a server behind them providing additional services. One such service is authentication. In most applications, there will be the need to identify users and provide additional services based on their rights, status, privacy, group, or any other category pertaining to the context of the application. A clear example of this is webmail applications, such as *Outlook* or *Gmail*.

Current web standards provide us with several options to perform asynchronous communications with a server. These are often called **AJAX** (*AJAX stands for Asynchronous JavaScript and XML*). In the most basic form, we could use the XMLHttpRequest object for these network communications, but the new specifications provide us with a direct function, fetch(), which is more convenient and standard between browsers. While these methods are perfectly valid, for other uses than simple needs, it is better to use a library that provides more functionalities built on top of these technologies—for example, one that provides an **API** to match **HTTP** request methods (GET, POST, PUT, OPTIONS, and DELETE) to easily consume **RESTful APIs** (where **REST** stands for **Representational State Transfer**, a type of architecture used in network communications). We will see more about this in *Chapter 8, Multithreading with Web Workers*. For now, just keep in mind that a library to handle network asynchronous communications is a better path. In our case, we will use the excellent **Axios** library (https://axios-http.com/), which you can install in your application with the following command:

```
$ npm install axios
```

Then, in your service or component, you can import and use the library with the following code:

```
import axios from "axios"
```

The library exposes methods to match each HTTP request (.get(), .post(), .put(), and so on), each one returning a promise that resolves to the result of the request or rejects it in the case of error.

With this introduction, we are ready to see some common patterns for authenticating users in our applications.

Simple username and password authentication

This is the simplest approach to authenticating users, where the validation of credentials is made by our implementation on the server. In this case, our server backend provides the API to validate a set of credentials, gathered by our SPA. Traditionally, the credentials are stored in the server, on a database, and the communication will be performed on top of **Secure Sockets Layer** (**SSL**) or encrypted communication, which are the same thing. Let's see the workflow graphically:

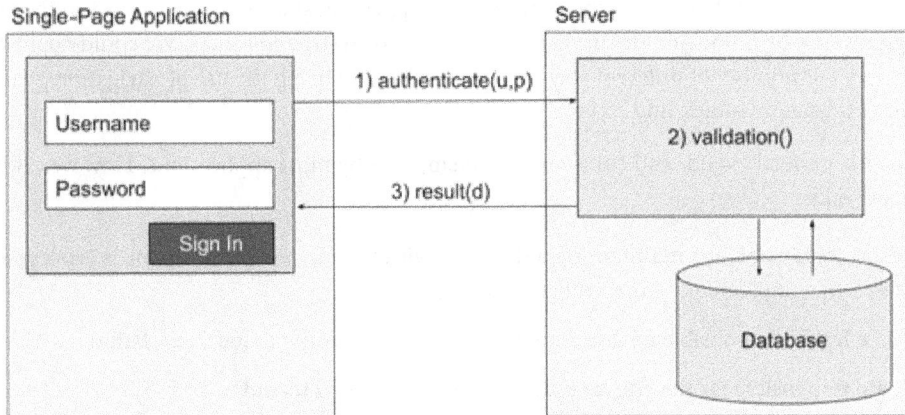

Figure 5.5 – Simple username and password authentication

In this workflow, the following occurs:

1. The SPA collects username and password values and transmits them to a specific endpoint in our server for authentication.

2. The server uses information stored in a database to validate the username and password.

3. The result of the operation is returned to the client SPA in response to their initial query (*1*).

Even though *Figure 5.5* shows the number of steps, consider that all this is done in just one network call and its reply. Developing the validation code on the server is beyond the scope of this book, but the code inside our service or Vue 3 component would look something similar to this:

```
import axios from "axios"
import {ref} from "vue"
const _username=ref(""), _password=ref("")
function doSignIn(){
axios.post("https://my_server_API_URL",
    {username:_username.value,password:_password.value})
  .then(response=>{
    console.log(response.status)
    console.log(response.data)
  }).catch(err=>{...})
}
```

As you can see, the implementation is quite straightforward and depends on our own logic and server API design. What is important to remember is to check the status of the response (everything between 200 and 299 is a success) and the data sent back by the server to act accordingly. Axios handles all the communication and data conversion for us (assuming our API receives and process JSON data).

In the case of success, we should save the result in our application state and allow access to the user accordingly, mostly by unlocking the navigation to private or restricted routes. We could apply this protection in a fair number of different ways, the most common being the use of navigation guards, the creation of dynamic routes, and so on.

This method is perfectly valid, and commonly implemented by most applications. However, it has several drawbacks:

- We are responsible for maintaining a database with usernames and passwords (encrypted, please!) and implementing the validation logic

- We are legally responsible for handling the user data according to local legislation

- We are responsible for the entire security of the system, end to end

- The user has to remember or be responsible for their own credentials

- We should provide ways to handle edge cases, as well as user problems and credential retrievals

These drawbacks are in no way a deterrent, but huge bullet points to keep in mind if we go this way. One way or another, most applications need to have a way to authenticate users, which depends on their own logic and implementation since not all of our users (depending on the context) will be willing to use another form of authentication, as we will see next.

OpenID and third-party authentication

Beyond security concerns, a major issue when dealing with authentication is how easily these credentials are lost or mishandled by the end user. This happens to us all. The more services we access online, the larger the number of credentials a user needs to "remember." There are many different methods to tackle this issue, to reduce the load on the user in keeping track of all these usernames and passwords. One such standard is the **OpenID** protocol (`https://openid.net/`).

The OpenID protocol authenticates users without the need to share credentials (usernames and passwords) between sites. It is based on the workflow of the **OAuth 2.0** protocol, which is used to securely shared information and resources without the need to use passwords as well. This is achieved by sharing tokens between the different actors. The standard for these communications is to use **JSON Web Tokens (JWTs)**. There is a lot to unravel in this paragraph, so let's see each one of these terms in a bit more detail so that we can better understand how this protocol works.

A JWT is a string that contains three sections, separated by a dot (`.`), and that have been encoded in Base64. Each section then encodes a JSON object with the following information:

- `Header`: This contains cryptographic information used to encode the token, such as the algorithm, the type of token (usually `JWT`), and in some cases even the type of data submitted in the payload.

- Payload: This object contains the information we want (need) to share, and is mostly "free format", meaning that it can contain any key:value pair as needed. However, there are a few well-defined fields that can also be used, such as "iat"(**Issued At Time**), which shares the timestamp for the creation of the token. Most importantly, this object must contain a unique ID for the user ("sub"field, for subject).

- Signature: The signature is a string form by concatenating the encrypted string representations of the header and payload, expressed in Base64. For the encryption, a secret key (a password) is used, only known to both the authenticating server and the website server.

When a website in the workflow receives a token, it decodes and validates it using the secret key, using the same method as the issuer. If the signatures don't match, then it is assumed that the token is corrupted or compromised, and it is rejected. A JWT can be intercepted and decoded by a third party, so this method acts as a failsafe against tampering. Let's see an example of the creation of a token:

- **Header**: {"alg": "HS256", "typ": "JWT"}. Here, we use the HS256 algorithm and declare the type used as JWT.

- **Payload**: {"sub":"1234567890","name":"Pablo D. Garaguso","iat": 1516239022}.

- **Secret encryption key**: secret key.

With the preceding information, a signature field is created with this formula (assuming we have a function that encrypts text using the HS256 algorithm):

```
HMACSHA256(base64UrlEncode(header) + "." + base64UrlEncode
(payload),"secret key")
```

Finally, the resulting strings in Base64 encoding are concatenated again to give us a perfectly functional token. Also, notice how each section (header, payload, and signature) appears separated by a period (.):

```
eyJhbGciOiJIUzI1NiIsInR5cCI6IkpXVCJ9.eyJzdWIiOiIxMjM0NTY3ODkwIiwi
bmFtZSI6IlBhYmxvIEQuIEdhcmFndXNvIiwiaWF0IjoxNTE2MjM5MDIyfQ.mPr551
xpsCgmIzp8EZuSCoy7t7iQNpp_iGzIR14E_Jo
```

To test this token, you can use a service such as https://jwt-decoder.com/. To validate it, however, you will need to use the secret key. You can test this at https://jwt.io, where you can also find more information about this standard.

In the OpenID protocol, JWTs are used to transmit and validate information between parties, hence why is so important to understand this concept well. There are several workflows recognized by the protocol. Let's see here a simplified representation of the **authorization code flow** (https://openid.net/specs/openid-connect-core-1_0.html) of the protocol with all the actors, and then see the parts we need to implement this in our Vue 3 SPAs:

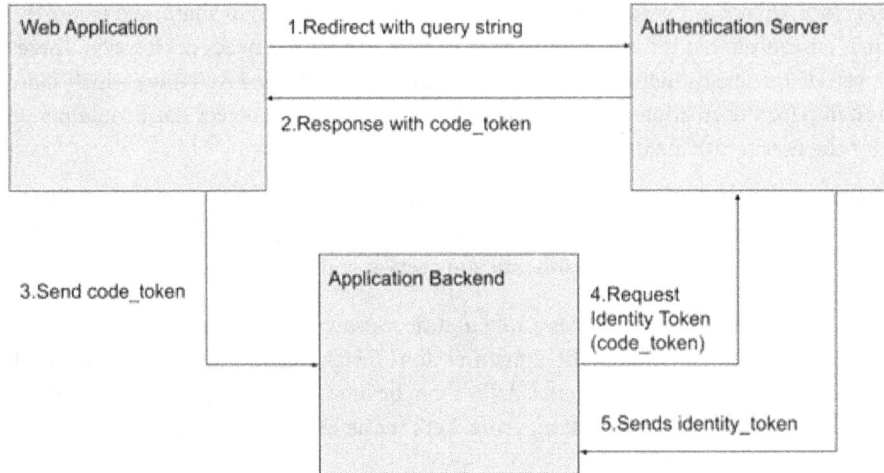

Figure 5.6 – The OpenID authorization code flow in all its beauty

As you can see, for this workflow to happen, we need three actors: 1) our SPA, handling multiple routes, 2) the authentication **service provider (SP)** server, and 3) our own backend server. It is possible to do the authentication and validation of our backend in the browser, then only needing two actors, but this is not recommended as it exposes the secret key in our JavaScript. The option is there, however, for embedded applications such as mobile apps where the user has no easy access to the page code (in hybrid applications).

In order to implement the workflow, the client (our application) must register with the authentication service. The process depends on each entity, but as a result, we will have registered the following:

- A `client_id` identification string, unique to our application.
- A `secret_key` value, which will be known only to the authentication server and our backend application. This will be used to encode and sign our tokens.
- A series of *endpoints* in the authentication server, and in our application where the user will be redirected in each step. Appropriate exchange of tokens will be done in these redirects as part of the query string in the URL.

So, let's see these steps in detail, and how to implement them in our Vue 3 application:

Step	Description

1	The user needs to be authenticated, so we redirect them to the endpoint given to us by the authentication server. The query string needs to be included with the following (mandatory) fields:
	
	• `scope`: **openid**
	• `response_type`: **code**
	• `client_id`: The client identification given by the authentication server
	• `redirect_uri`: The same address that we registered with the server where the user will be redirected upon successful authentication
	• `state`: Any data or application state that we want to receive in return after the authentication
	
	To prepare the redirection URL, we first create an object with the preceding fields and values, and use the `URLSearchParams` creator to create a query string (see `https://developer.mozilla.org/en-US/docs/Web/API/URLSearchParams`):
	
	`const query_data={scope:"openid", ...},`
	`query_string=new URLSearchParams(query_data).toString()`
	Next, we can use the location object to execute the redirection:
	`location.assign("https://auth_endpoint" + "?" + query_string)`

2	On successful authentication, the authentication server will redirect the user to the endpoint that we registered as the receiver. The query parameters sent will depend on the result of the operation: • Successful sign-in: ▪ `code`: The **code_token** that needs to be exchanged later for an `identity_token`. ▪ `state`: Any data we sent to the server and want back. We can use this to redirect the user inside our application, for example. • Unsuccessful sign-in: ▪ `error`: An error code as specified by the protocol (`interaction_required`, `invalid_request_uri`, and so on).

Step	Description
	The redirect will trigger our application being loaded, and the router will render our designated component. In our script setup, we need to capture the query string passed to us, to be used later in the next step. One approach to do this without the use of third-party libraries is with the following code:

```
import {useRoute} from "vue-router"
const $route=useRoute()
if($route.query.error){
    // The authentication failed, take action
}else{
    // Authentication succeeded do something
    sendToServer($route.query.code)
}
```

Step	Description
3	In this step, we just send the code received to our backend, which would mean implementing the `sendToServer()` function mentioned previously. Since now we are dealing with our own implementation, the way to do this is trivial. In this example, we are using Axios:

```
import{axios}from "axios"
function sendToServer(code){
axios
  .post("our server URL", {code})
  .then(result=>{
    // Set the token in our headers
    axios.defaults.headers.common={
    "Authorization":"Bearer " + result.data.identity_token
    }
  }).catch(()=>{
    // Handle the error
})}
```

In the previous example, we have sent to our server the code_token string and received from our server the identity_token string as response. We then go one step beyond and set the default headers for our application to use the standard Authorization header, with a Bearer token. From then on, our server only needs to check the headers and verify that the operation requested belongs to a valid user.

Implementing the validation of the tokens and *steps 4* and *5* go beyond the scope of this book, as we are focusing on Vue 3 applications. As you can see, the part that our SPA needs to handle is quite simple and does not really involve much code (some error checking was omitted for the sake of brevity).

There is a good number of syndicated authentication services, both free and paid, that we can implement in our application. Most common these days is to see badges redirecting users to use them, such as signing in with *Google*, *Facebook*, *Twitter*, *GitHub*, *Microsoft*, and so on. There are also meta-services that provide all the aforementioned providers inside well-packaged libraries, such as **Auth0** (https://auth0.com/, now part of **Okta**, https://www.okta.com/). When it comes to implementing this workflow, we are certainly not short of options.

Passwordless or one-time password (OTP) authentication

Another solution to remove the use of credentials is to remove them altogether with passwordless access. The basic idea is to rely on the security of another system (email, mobile texts, authenticator apps, and so on) to validate the user. The process generates a time-sensitive "one-time use only" code and sends it to the user through the supporting system via the backend service. The frontend service (the application) awaits the right to be entered by the user in a determined time frame. For example, a common implementation is for the backend to send a text message to the user's phone containing the code, which has to be entered into the application before the time expires.

Here is a visual representation of this workflow, considering that the user has been registered with an email or phone number. These are supposed to be *well known*, meaning that the ownership has been verified:

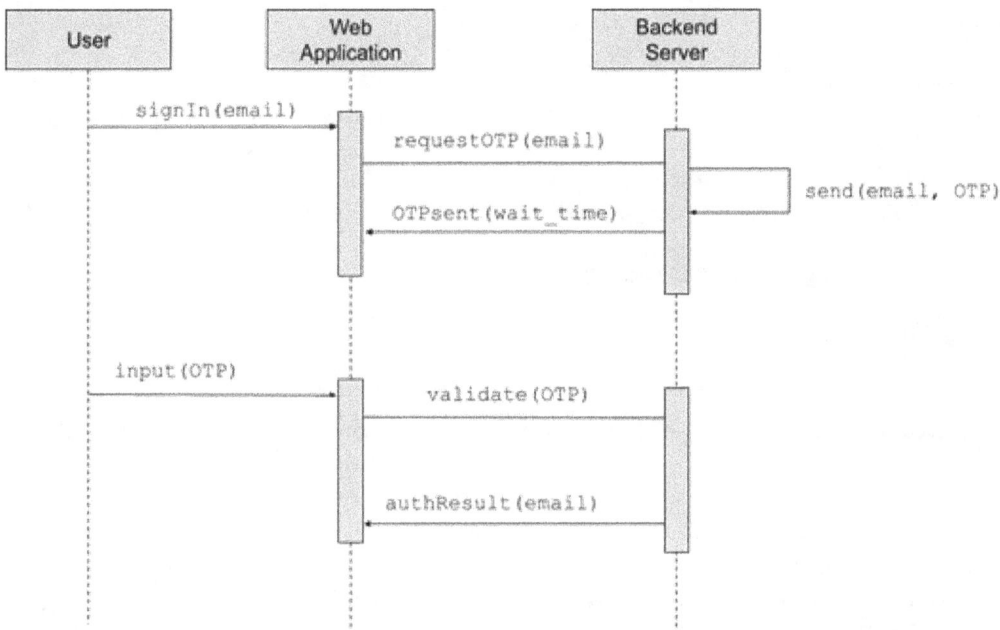

Figure 5.7 – Passwordless authentication based on email

In the preceding workflow, notice that the OTP code never reaches the web application until the user inputs it. The validation occurs in the backend, not in the frontend. This makes our application very simple, as it only needs to collect the email first and submit it to the server, and then wait for the specified time for a new input. In a service or component, using Axios, this code would look something like this:

```
const _user_email=ref(""),
       _wait_seconds=ref(0),
       _show_input_code=ref(false),
       _otp_code=ref("")
function signInUser(){
    axios.post("https://requestOTP_url",
                {email:_user_email.value})
    .then(result=>{
        _wait_seconds.value=result.data.wait_time;
        _show_input_code.value=true;
        startOTPtimer();
    }).catch(err=>{...})
}
function startOTPtimer(){
    let interval_id=setInterval((()=>{
    if(_wait_seconds.value>0){_wait_seconds.value--;}
    else{clearInterval(interval_id);}},1000)
}
function checkOTP(){
    axios.post("https://validateOTP_URL",{code:_otp_code.value})
    .then(result=>{
      if(result.status>200 && result.status<300){
        // User validated, proceed to protected route
      }else{
        // Validation failed. Redirect to error route
      }
    }).catch(err=>{...})
}
```

In the preceding code, we have omitted the imports and the template, as at this point they should be trivial for the reader. Our template should have at least an input to collect the email of the user, and a second input to collect the OTP code, plus two buttons to trigger on click the signInUser() function and the checkOTP() function. The first one will pass the email to the backend, and wait for a reply with at least a time in seconds to wait, which we use to start a timer (it is always good to let the user know how much time they have to enter the code). Nowadays, for emails and text messages, the standard is 60 seconds. When this happens, we also hide the first input and then show the "OTP" input form. When the user enters the code and clicks **Submit**, the checkOTP() function is activated, and we pass the code again to the server waiting for a reply. On success, we can redirect the user to a protected area according to our application logic. Considering the triviality of the template, it would be a good exercise for the reader to create the component and template by themself. Then, a possible solution can be found in the code examples, in the `chapter 5` folder.

Following a progressive approach to security, the next step is to join the previous approaches into a common new process: **two-factor authentication** (**2FA**), which we will see now.

2FA - Two Factors Authentication

In 2FA, our application merges two or more of the previous approaches to validate a user. The key concept behind this method is that even a third party or simple username and password are not enough, and the user needs to have a "secondary factor" to be validated—for example, the use of a registered email, phone number (for SMS submission of codes), authentication apps (example: Google Authenticator), a USB device, a security card (with a chip or band reader), and so on.

The workflow is simple but does require more from our backend than from our frontend application. Once our SPA authenticates our user using any of the methods listed previously, a second request is triggered on the backend to submit the proper query to the security device. Let's assume that our user receives an SMS from our server with a code. Our SPA will wait and collect this code during a specific time frame (usually, 60 seconds), and submit it to the backend to a specific endpoint. It is the server that then validates the code. In reality, this is like having two or multiple passwords, whose validation is made in cascade. Any step fails, and the entire operation is discarded.

Here's a visualization of the process:

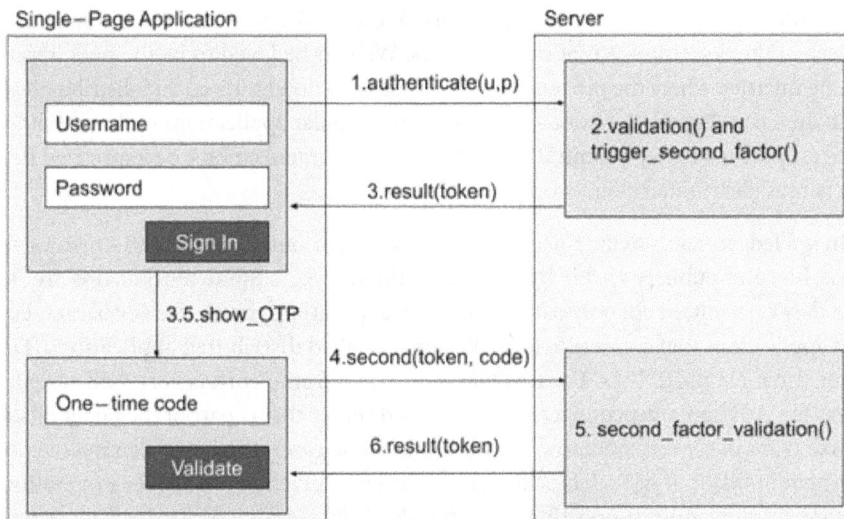

Figure 5.8 – A simplified view of our SPA and server interactions with 2FA

As we can see from the simplified workflow, the process of validating a user with 2FA (as with passwordless and OTM methods) does not rely so much on the code or some specific cryptography but on the use of clever communication and data isolation. The data and validation process never leaves our server or could be visible to an end user, even if opening the code of our SPA. In a way, you can think of this workflow as the concatenation of an OpenID or credentials authentication, followed by an OTP implementation.

Our application's main responsibility is just to collect the bits of data that make up each step and relay them to the server. In between, we could change the route or update the interface, but this implementation is trivial, so we won't see specific code here (you can see previously how to programmatically change a route, for example).

In general, 2FA is considered a "more secure method", but it is not without its drawbacks, and it may not be the right fit for every application. For example, what happens if you know your username and password, but lose your secondary device (your mobile gets stolen, hacked, and so on)? Organizations that use this method often provide a way to recover your identity, often with expensive implementations (think of a bank and phone service). In the end, this method does apply one more layer of complexity in the authentication of users, and with it, another possible point of failure, ending with very frustrated users if not handled properly.

Let's see next another authentication method that is gaining traction as the new kid on the block of authentication patterns: Web3 authentication.

Web3 authentication

Before we dig into the topic, we need to define what **Web3** is. There seems to be some confusion as to the extent of the definition, so for our purposes, Web3 is believed to be the next iteration or evolution of the internet, where the processing power is done in decentralized and distributed servers, using blockchain technologies. The most well-known and popular applications of these technologies nowadays are cryptocurrencies, decentralized self-governing organizations, decentralized finances, play-to-earn games, distributed cloud storage, and much more.

A **blockchain** is a ledger that is maintained by a network of distributed computers. Anything written to it is immutable, and publicly visible by anyone on the network. Some blockchains are "smart", meaning that they can contain not only data but also run applications, much like any backend service. The frontend applications that connect to a blockchain are called **distributed applications (DApps)**, which arc, for the most part, *SPAs*. For this task, the Vue 3 framework is very well suited, as we have seen thus far. A DApp must connect with a backend server that is part of the target blockchain network. These types of servers are known as **nodes**. In some cases, the DApp can interact directly with the blockchain. Most, if not all, blockchains use cryptocurrencies to regulate operations and reward the contributing nodes supporting the network. Cryptocurrencies are logically assigned to a unique blockchain ID called a "wallet". These wallets implement some very smart cryptographic techniques to validate each other when performing operations, through the use of public and private keys. A user may have many wallets. There are no emails or ways to recover lost keys in a blockchain, and each wallet is unique.

In order to resolve all this cryptographic signage and validation, and to make it easier for users, there are special plugins for browsers called "digital wallets," as well as mobile application wallets that also implement web browsing. These applications hold the credentials and do the heavy lifting when dealing with the blockchain. There are, of course, numerous libraries to do the same tasks in pure JavaScript, but this goes beyond the scope of this book. What we will see next is how in our SPA, we can leverage the power of these technologies to identify a user, even automatically when visiting our application page.

We will focus mainly on the biggest smart blockchain, the Ethereum network, as an example. The same workflow with more or fewer steps applies to other networks using different SDKs, so the migration or incorporation of additional blockchains is not too far away from our examples. The basic conceptual workflow is as follows:

Import a library to connect to the network in our JavaScript, either through a library such as `web3js` (`https://www.npmjs.com/package/web3`), `ethjs` (`https://www.npmjs.com/package/ethjs`), or use the one injected directly by a browser wallet—in our example, **MetaMask**, in `window.ethereum`

- Using the `ethereum` object, we request the user to connect their wallet to our site and retrieve the selected wallet address

- Our application then can send to our backend the wallet ID (that is public) and use it as the unique ID for the user's account

As just mentioned, we will use the object injected by **MetaMask** (`https://metamask.io/`) since it is one of the best-known browser wallets. In this case, here is the code that requests the current user's wallet:

```
ethereum
.request({ method: 'eth_requestAccounts' })
.then(
    result=>console.log(result[0]),
    err=>console.log(err)
)
```

That's it! The highlighted line prompts *MetaMask* to open a new window and request the user's permission to connect their wallet to your web application, and then return a handy promise. If approved, the result will be an array of strings, where the first position is the wallet address for the current network. If rejected, an error will be triggered.

> **Tip**
>
> With MetaMask, you can open the **Developer Tools** in the browser and type in one line the preceding code to test it.

Using **MetaMask**, the same code for the **Ethereum** network also works when connected to the **Polygon** and **Binance Smart Chain** networks (three for the price of one!). Other networks and wallets, such as the Phantom wallet, follow the same principle and inject into the `windows` object a new object called `.solana`. Check the documentation of your target blockchain to become acquainted with the details of each implementation.

Interacting with each blockchain and the code thereof is beyond the scope of this book, so we will limit ourselves to identifying the user by their wallet address. Having obtained this identification, it is up to our application logic to store them for future reference, as it acts like the user ID.

There are also third-party solutions to authenticate and interact with multiple blockchains, and we should consider them before implementing our own solution.

Summary

In this chapter, we considerably improved our application and created a sound SPA with navigation using the Vue router. This is an important concept to segment our application and organize the work between the members of a development team. Fractioning our application following the navigation path makes development and maintenance easier to approach and better organized. We also learned several authentication standard patterns that we can consider for our applications, covering a good number of scenarios used today in the industry, from the very basic username and password, all the way to the new Web3 DApps. We also took time to understand how standard protocols such as OAuth work, as well as OTPs, and how these can be implemented for an extra layer of security as a second-factor authentication. All these skills are relevant and necessary for today's web application standards.

In the next chapter, we continue expanding our technical knowledge with the introduction of **progressive web applications** (PWAs).

Review questions

We have covered multiple different topics and introduced new concepts in this chapter. Use the following questions to solidify what you just learned:

- When is better to use a SPA instead of an MPA and vice versa?

- What are the benefits of using a router in our SPA? Name at least three from your own analysis.

- How can you use views to define the layout of your application?

- How can you access the parameters and query string passed to your route in your JavaScript?
- What are some common standard patterns to authenticate users?
- What are some security considerations when authenticating users in a SPA?

6

Progressive Web Applications

In this chapter, we will see the next evolutionary step for web applications: **progressive web applications** (**PWAs**). This term may not seem descriptive enough, but it refers to a group of technologies that create the general concept and can be implemented gradually or partially. The basic idea behind it is to bring a web application out of the context of the browser and implement it in any type of device, to act and behave as much as possible to a native application. This is done thanks to the implementation of new APIs in the browser engines, as well as integrations among the most popular operating systems for desktop and mobile devices. The starting point for a PWA is, of course, a **single-page application** (**SPA**).

By the end of this chapter, we will have learned the following:

- What makes a SPA a PWA, and which technologies are involved
- How to implement manually a responsive SPA, manifest file, service workers, offline storage, and so on
- What *service workers* are
- How to use Vite plugins to automate the creation of PWAs
- How to test the readiness of your application using *Google Lighthouse*

From the preceding list, we will concentrate on learning the scaffolding for several technologies, setting the foundation to use them later, implemented in detail in *Chapter 7, Data Flow Management*, and *Chapter 8, Multithreading with Web Workers*. By the end of these chapters, you will know how to create PWAs that make good use of today's computing power, making them responsive, reliable, and performant.

Technical requirements

In order to follow along with this chapter, you will need the code examples found in the repository at https://github.com/PacktPublishing/Vue.js-3-Design-Patterns-and-Best-Practices/tree/main/Chapter06. The text code examples in this section may not be enough to create a working example, without the additional code from the repository.

Check out the following video to see the Code in Action: https://packt.link/SBZys

PWAs or installable SPAs

A PWA is not a single setting or technology, but a systematic enhancement of a web application to comply with certain conditions, be it either a **multi-page application** (**MPA**) or a SPA. However, they really shine and come to life when these technologies are applied to SPAs, giving us powerful applications that blend the line between online/offline and desktop or web. The term **progressive** used here has the same connotation as we have discussed previously when applied to the Vue framework—an incremental application of web technologies.

PWAs are then treated somehow specially by the browsers and the operating system. They can be installed alongside native or desktop applications and manage network communications (to send, receive, cache files, and even push notifications from the server). At this point, it is important to note that we are no longer referring only to desktop computers but also to mobile devices such as tablets and phones, and different operating systems. It is because of this multiplatform approach that special consideration needs to be taken if the intention is to cover a user base on different devices, such as the use of special and dedicated CSS rules to adapt the UI to different sizes (the so-called **responsive applications**), different icons, and colors to blend with local user customizations at the operating system level (for example, light and dark modes), and so on. Moreover, PWAs have the capacity (just as with SPAs) to store content for offline use and, hopefully, should also provide some functionality for offline use. To accomplish all of this, at the bare minimum, a PWA must comply with the following requirements:

- The web application must be served through a secure connection (HTTPS).
- The application must provide a manifest file.
- It must provide and install a service worker.

When all these conditions are met, the browser or operating system may prompt the user to "install" the application. If the user accepts, the manifest file will be used to customize the appearance of the application to match the local operating system (icons, names, colors, and so on), and it will appear alongside all the other applications in the system. When run, it will open in its own window (if so selected) outside the confines of the web browser, just as with a regular native application. Internally, it will still run over the browser engine using web technologies, but the intent is that this will be transparent for the user, providing the best of both worlds. Chances are that a user may have been using PWAs instead of regular applications without knowing. Successful examples of this approach are Starbucks, Trivago, and Tinder (`https://medium.com/@addyosmani/a-tinder-progressive-web-app-performance-case-study-78919d98ece0`).

This creates a good number of advantages that overpower the complications of creating a web application to match the different installation scenarios:

- One single code base to install an application on multiple devices (desktop, mobile, …) and operating systems (Windows, Linux, macOS, Android, iOS, and so on)
- Support push notifications from the server, manual handling of caching, offline use, and so on

- They integrate with the local operating system

- Updates are transparent for the user and are much faster than a traditional application (for the most part)

- Developing a PWA incurs much fewer costs than the equivalent targeted individual applications for each platform

- One can use all available web technologies, frameworks, and libraries

- Can be indexed by search engines, and the distribution and installation do not depend on proprietary application stores

- They are responsive, safe, and fast, and can be shared with just a link

- You can access local devices using standard web APIs, such as the local filesystem and USB devices, use hardware accelerated graphics, and so on

- Some proprietary application stores allow you to re-package your PWA and distribute it as a regular application (Microsoft Store, Amazon Store, Android Store, and so on)

There are more advantages, but these may be enough to make our case for them. Also, it is easier to add the necessary elements to our SPA to make it a PWA. This may make PWAs look like the silver bullet of applications; however, there are a few caveats and drawbacks to consider as well:

- The performance of a PWA is good but will always fall behind a native application for certain specific scenarios. The same may happen in older hardware—they will run, but performance may suffer.

- Apple devices fall a little behind in the adoption of some web technologies or limit them purposely for PWAs (for example, server push notifications).

- There's a need to dedicate a bit more effort to cover different user experience scenarios on multiple devices (but slightly more than for a normal responsive web application).

- Some application stores will not allow PWAs (specifically at the time of this writing, the Apple App Store). Also, the application will not benefit from the exposure and *foot traffic* from an app store.

Overall, the advantages outweigh considerably the disadvantages. As web technologies continue to advance, PWAs benefit more from them and become more ubiquitous. Now, with a better understanding of what a PWA is and what it can do, let's upgrade our SPAs into PWAs.

Upscaling a SPA into a PWA

The first requirement mentioned previously is to serve the application over a secure connection. We will see how to accomplish this by installing a free SSL certificate in our server using **Let's Encrypt**, in *Chapter 10, Deploying Your Application*. With that in mind, let's see how to fulfil the other requirements.

The manifest file

Adding a manifest file is the starting point to turn our application into a PWA. It is nothing less than a JSON file with well-known fields that instruct the browser or operating system on how the application should be installed on the desktop or mobile device. This file must appear linked in the `head` section of our `index.html` file, and while it could be arbitrarily named, the convention is to use the name `manifest.json` or `app.webmanifest`. The official specification suggests the `.webmanifest` extension, but at the same type clarifies that the name is not really important as long as the file is received properly with the `application/manifest+json` - **Multipurpose Internet Mail Extensions** (**MIME**) type (see `https://www.w3.org/TR/appmanifest/`, section *1.1.2*). In our code examples, we will use the name `manifest.json` for simplicity:

```
<link rel="manifest" href="/manifest.json">
```

Notice from the previous code that the file is placed at the root of our application, and the `rel` attribute must be `manifest`. The field attributes in our manifest file can appear in any order, and all of them are considered *optional* by the aforementioned specification. However, some platforms do expect a minimum set of attributes that we will consider *necessary*. The common practice also demands other attributes that we will catalogue as *recommended*, and finally, some attributes in the specification are used often in app stores, social media, and so on to present or describe the application, so we will refer to these as *descriptive* fields. This classification is not part of the specifications but is useful to guide you in the implementation. Here is a list of the most common and useful attributes:

Classification	Attribute
Necessary	
short_name	A short name to be used when there is not enough space to display the entire name of the application. In mobile devices, it is often used for the icon name.
name	The full name of the application.
icons	An array of objects, each one representing an individual icon to be used in different contexts. Each object has at least two attributes: • `src`: The path to the image • `sizes`: A string with the dimensions of the image
start_url	The URL where the application should start, as set by the developer.

display	A string that represents how the application is presented: • `fullscreen`: Fullscreen, but showing browser UI. • `standalone`: Like `fullscreen`, but without browser controls. On a desktop, windows controls will still display. • `minimal-ui`: Like `standalone`, but with basic navigation to move forward and backward, print, share, and so on. • `browser`: The application is open in the default browser.
Recommended	
theme_color	A string representing a CSS color for the application. It is upon the OS to decide how to use this value (usually, applied in the window title bar).
background_color	A string that represents the background color of the application when it is launched and before the actual styles from the application are applied.
orientation	Mostly used in mobile devices, it defines the orientation that the application must use—for example, `landscape`, `portrait`, `any`, and so on.
lang	A string that defines the main language of the application.
Descriptive	
shortcuts	This is an array of objects that define direct menu options for tight integration with the operating system. Usually, these appear in the context menu, such as when a user right-clicks on the application icon. Each shortcut object must contain at least `name` and URL, and—optionally—a `description` and `icons` array.
description	A string with a short description of the application.
screenshots	An array of objects, with the following fields: • `src`: URL for the image • `type`: MIME type of the image • `sizes`: A string with the dimensions of the image

Table 6.1 – Manifest fields

In practice, I would recommend that the necessary and recommended fields are completed for each PWA, while the descriptive fields are used as needed based on the context of your application. Additionally, research your target platforms for additional supported fields that are not part of the standard specification.

Following the preceding table, here is an example of a `manifest.json` file:

```
{
    "short_name":"PWA Example",
    "name": "Chapter 6: Progressive Web Application Example",
    "start_url":"/",
    "display": "standalone",
    "theme_color":"#2979FF",
    "background_color":"#000",
    "orientation": "portrait"
}
```

As you can see, creating a manifest file is not much additional effort and is an easy addition to our SPA.

Testing your manifest

Once you have created your manifest file and linked it to your `index.html` file, you can use the Developer Tools in a browser to check that it has been properly loaded. For example, when using Google Chrome, in the **Application** menu, we can see here that the example file has been properly loaded:

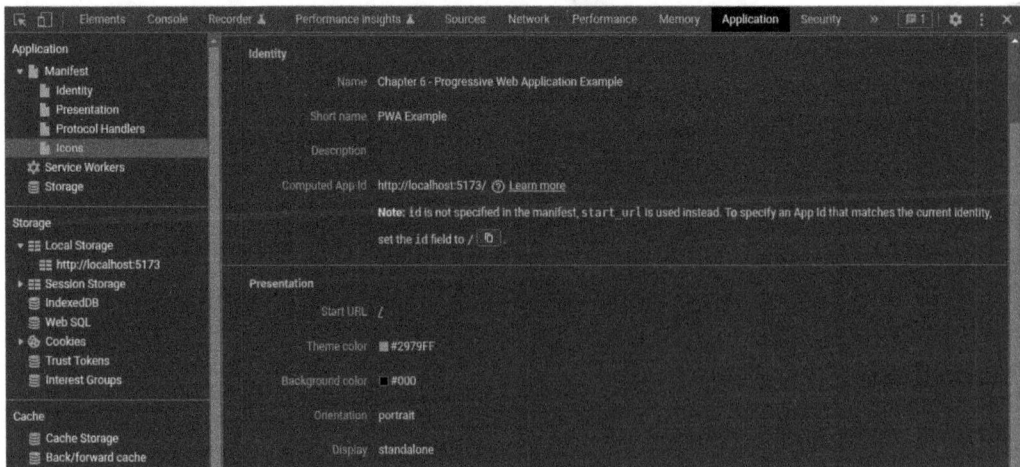

Figure 6.1 – Developer Tools on Google Chrome, showing the manifest file

However, there is one more topic related to the installation of the application that we must review: when and how does the user know that the web application can be installed? This is where the *Install prompt* comes into play, which we will see next.

Install prompt

Each platform (mobile or desktop) has its own method of determining when a PWA that meets the installation criteria can be installed. This could trigger a notification for the user to accept the installation after a certain amount of time, or only provide a UI to do so. On mobile devices, the installed PWA will be placed on the home screen alongside other native applications, while on a desktop, it may be placed inside the browser and/or also in the main menu. Also, in mobile operating systems such as Android, a splash screen will be automatically created with the theme and background colors and the application icon provided in the manifest. Independently of how and when the PWA can be installed, it is important to know that it can only be done with the user's consent and initiative. We cannot trigger the installation automatically from the code without the user's approval.

The basic installation flow is as follows:

1. When the platform detects that our application can be installed, it will trigger an event in the window object called `beforeinstallprompt`. We can cache this event to trigger the prompt later from our application.

2. The user initiates the install either through the platform UI or through our PWA-provided method (such as a button).

3. The platform will prompt the user to accept or reject the installation.

4. If the user accepts, it will install the PWA and trigger another event named `appinstalled`.

This is a rather simple workflow. However, the `beforeinstallprompt` event is triggered only once, so if the user rejects the installation, we need to wait until the browser triggers the event again.

Now that we understand how things will work, it is time to see this in code. Consider that in our Vue 3 component's template, we have the following elements:

```
<p v-show="_install_ready && !_app_installed">
    Install this app
    <button @click="installPWA()">Install</button>
</p>
<p v-show="_app_installed">
    Progressive Web Application installed
</p>
```

As you can see, we have two paragraphs that will show according to the value of the `_install_ready` and `_app_installed` reactive variables, both Boolean. The first will appear when the PWA is ready to be installed and will provide a button to trigger the installation through the `installPWA()` function. The second will show once it has been performed.

Our code in the script section is also rather straightforward:

```
import { onMounted, ref, onBeforeUnmount } from 'vue'

const
    _install_ready=ref(false),
    _install_prompt=ref(null),
    _app_installed=ref(false)

// Detect PWA installable
onMounted(()=>{
    window.addEventListener("beforeinstallprompt",savePrompt)
    window.addEventListener("appinstalled",handleAppInstalled)})

function savePrompt(event){
    event.preventDefault(); // Prevents mobile prompt
    // Save reference to the event, to activate it later
    _install_prompt.value=event;
    // Notify UI that the application can be installed
    _install_ready.value=true;
}

function installPWA(){
    // Trigger the installation prompt
    if(_install_prompt.value){
        _install_prompt.value.prompt()
    }
}

function handleAppInstalled(){
    _install_prompt.value=null;
    _app_installed.value=true;
}
```

In the previous code, we register two listeners when our component is mounted on the page, one to manage and cache the installation prompt, and another to detect when the application has been installed. Some parts have been omitted to keep the code simple, but the full component with styles can be found in the GitHub repository.

While the preceding example is rather simplistic, there are some well-known patterns to promote or introduce the installation option to the end user. They all rely on the same logic of capturing the event and prompting it later while showing the trigger element. The implementation is trivial and has more to do with design than a coding pattern, so we will only see the mock-ups here:

- Simple **Install** button (as in our example application):

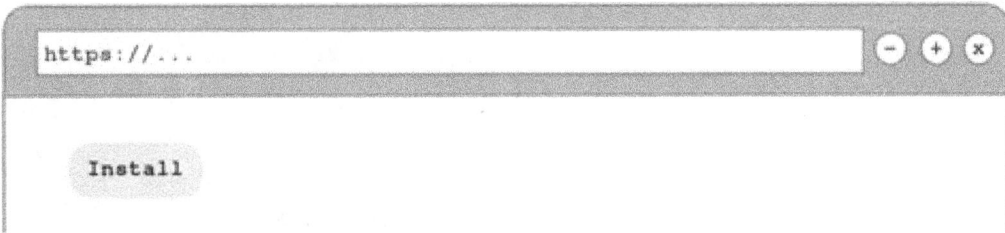

Figure 6.2 – Simple Install button

- Menu **Install** button—placed in the main navigation:

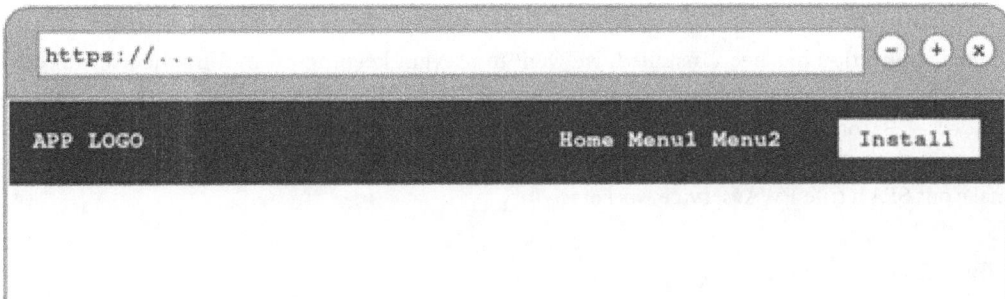

Figure 6.3 – Main menu Install button

- An overlay notification:

Figure 6.4 – Overlay notification

- An on-top overlayed element, such as an installation banner (either before the header or at the bottom of the viewport):

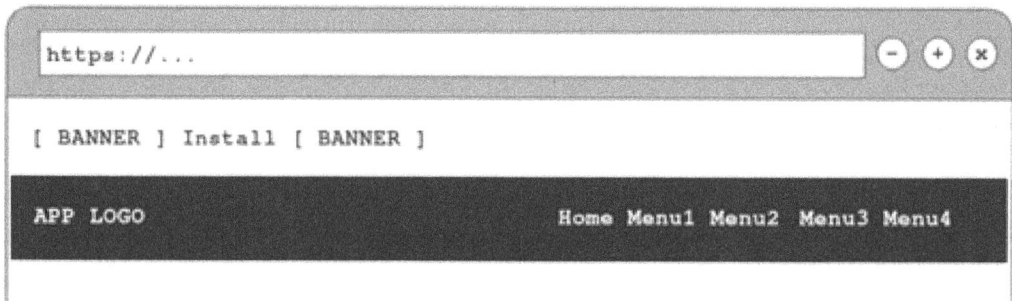

Figure 6.5 – Installation prompt banner

Once the application has been installed, we want to prevent keeping prompting the user for an install. In this case, it is recommendable that we save the offline flag, in `localStorage`, a cookie, on `indexeDB,` or mark the start URL of our application to a specific location. We will see offline persistent storage options in *Chapter 7, Data Flow Management*. Now, it's time to look at the last item to make our SPA a true PWA: service workers.

Service workers

A service worker is a JavaScript script that runs on a separate thread, as a background process to your application. It acts like a proxy for the network, intercepting all calls and behaving according to a programmed strategy to serve pages and data.

We can have multiple service workers as each one is responsible for their scope. The scope is defined as the directory (URL path) where the source file for the service worker is located. Thus, a service worker placed at the root of the application will handle the entire SPA/PWA.

Service workers are installed without user intervention, so they can be used even if the user does not install the PWA. They have a well-defined life cycle (see `https://web.dev/service-worker-lifecycle/`), triggering events for each accomplished state. To start, a service worker needs to be first *registered*, then it becomes *activated*, and eventually, we can also *unregister* it. Once the service worker is activated, it will not take control of the application communication until the next time the site is accessed.

The most common strategies to program a service worker are as follows:

- Serve the cache only
- Serve network only
- Try to serve the cache first, fall back to the network

- Try to serve the network first, fall back to the cache

- Serve the cache first, update the cache next

When considering the cache and offline strategies, we need to consider what are the files and assets that our application needs to run that will have little or no change, to cache them. We also need to identify routes that should never be cached.

To use a service worker, we register it in our main.js file with the following lines:

```
if(navigator.serviceWorker){
    navigator.serviceWorker.register("/service_worker.js")
}
```

In these lines, we first test if the current browser has capabilities to use service workers, and if so, we register it. As we can see, we have placed the worker at the root. For this example, we will use a cache-first, network-fallback strategy manually for all network calls:

```
// Set strategy, cache first, then network
const CACHE_NAME="MyCache"

self.addEventListener("fetch", event=>{
    // Intercepts the event to respond
    event.respondWith((async ()=>{
    // Try to find the request in the cache
    const found=await caches.match(event.request);
    if(found){
        return found;
    }else{
        // Not cached fount, fall back to the network
        const response=await fetch(event.request);
        // Open the cache
        const cache=await caches.open(CACHE_NAME);
        // Place the network response in the cache
        cache.put(event.request, response.clone());
        // Return the response
        return response;
    }
    })())
})
```

The previous code is based almost verbose on the example provided in the **Mozilla Developer Network** documentation at https://developer.mozilla.org/en-US/docs/Web/Progressive_web_apps/Offline_Service_workers. The comments in the code will help you understand the logic for the implementation of the strategy. However, using the basic APIs available to the service worker can be cumbersome, if not verbose. Instead, it is more convenient to use a framework or library to handle them and implement more complex strategies. The standard today is to use **Workbox**, made by **Google** (`https://developer.chrome.com/docs/workbox/`). We will not use it directly, but through a Vite plugin that we will see in the next section.

With all the code seen thus far, our PWA is ready to work and be installed. If we run the example application in the development server, we will notice that it can be installed. Using either the browser UI or our **Install** button, we will receive the following prompt:

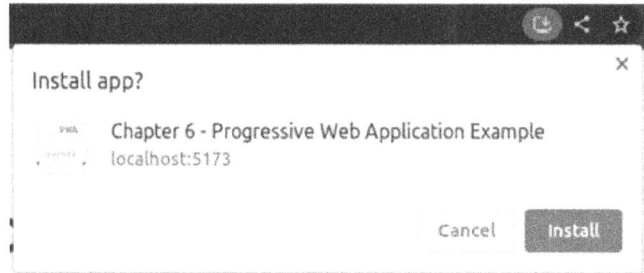

Figure 6.6 – PWA install prompt from localhost

Adapting manually our SPA to turn it into a PWA is not complicated, but it does require some manual work. However, with the choice of tools we have, we can do better. There is an easier way to generate and inject both the manifest file and service worker as part of our workflow directly into our SPA: using a Vite plugin.

Vite-PWA plugin

In the Vite ecosystem of plugins, there is an excellent zero-configuration Vite-PWA plugin (`https://vite-pwa-org.netlify.app/`). Out of the box, it provides us with great functionality without much manual work. We install the plugin as a developer dependency with the following command in the terminal:

```
$ npm install --save-dev vite-plugin-pwa
```

Once it has been installed, we must register it in the Vite configuration. Modify the `vite.config.js` file to match the following:

```
import { defineConfig } from 'vite'
import vue from '@vitejs/plugin-vue'
import { VitePWA } from 'vite-plugin-pwa'


export default defineConfig({
plugins: [
```

```
vue(),
VitePWA({
    registerType: "autoUpdate",
    injectRegister: 'auto',
    devOptions: { enabled:true },
    workbox: {
        globPatterns: ['**/*.{js,css,html,ico,png,svg}']
    },
    includeAssets:
              ['fonts/*.ttf','images/*.png','css/*.css'],
    manifest: {
        "short_name": "PWA Example",
        "name": "Chapter 6 - Progressive Web Application Example",
        "start_url": "/",
        "display": "standalone",
        "theme_color": "#333333",
        "background_color": "#000000",
        "orientation": "portrait",
        "icons": [
            {
              "src": "/images/chapter_6_icon_192x192.png",
              "sizes": "192x192",
              "type": "image/png"
            },
            {
              "src": "/images/chapter_6_icon.png",
              "sizes": "512x512",
              "type": "image/png"
            },
            {
              "src": "/images/chapter_6_icon.png",
              "sizes": "512x512",
              "type": "image/png",
              "purpose":"maskable"
            }
        ],
        "prefer_related_applications": false
    }
})]
})
```

Using this plugin, we unload the burden of generating the service worker and web manifest to the bundler. This is necessary since with each production build, Vite will generate different filenames for each script according to our strategy to lazy load components, as we discussed in the previous chapter.

In the preceding example, we pass into the `VitePWA()` plugin an object with some sensible options for the automatic creation and injection of our manifest and worker. If we need finer control over the service worker strategy created, as well as with the web manifest, it is possible to use the plugin in "inject mode" and provide a base file for our service worker. In this case, the script will be injected with the generated files from the build process. Underneath, this plugin uses **Workbox**, a tool that we mentioned before and that we can customize directly through the `workbox` field. Going further into the details of the different implementations and strategies goes beyond the scope of this book, but the reader should consult the documentation about the **Vite-PWA** plugin and **Workbox** for specific contexts and use cases.

Testing your PWA score with Google Lighthouse

Chrome-based browsers provide together with the developer tools a utility called Lighthouse, specifically designed to test and rate web pages, and the readiness of PWAs. To access this tool, after you have opened your application in the browser, follow these steps:

1. Open the developer tools (by pressing *F12* in Windows/Linux, *Fn + F12* in Mac, or through the browser menu).

2. Click the **Lighthouse** menu toward the further right.

3. Select **Mobile** or **Desktop**, plus make sure the **Progressive Web App** category is checked.

4. Click **Analyze page load** in the top-right corner of the tool.

The developer tools should look something like this:

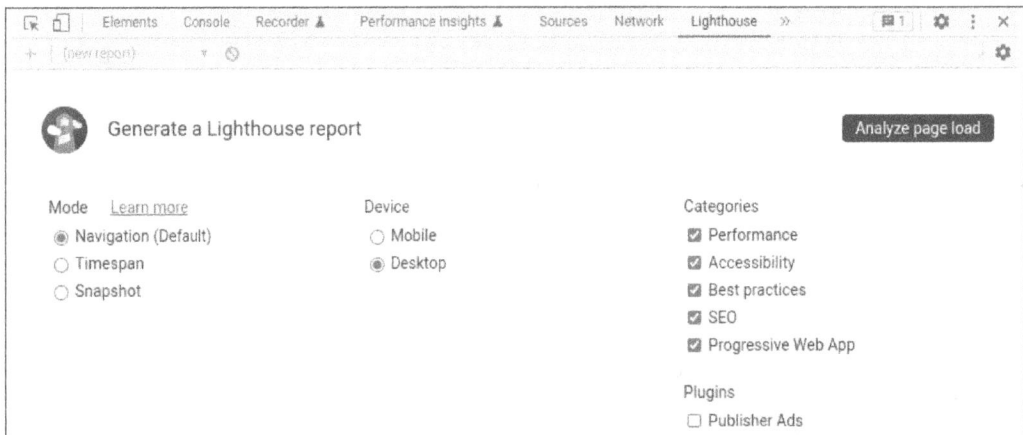

Figure 6.7 – Lighthouse utility

The tool will run a number of tests, and each different category will display a rating, as well as a detailed list of items that have either passed or failed. If our application does not qualify to be a PWA, the items marked in red will tell us why and how to fix them:

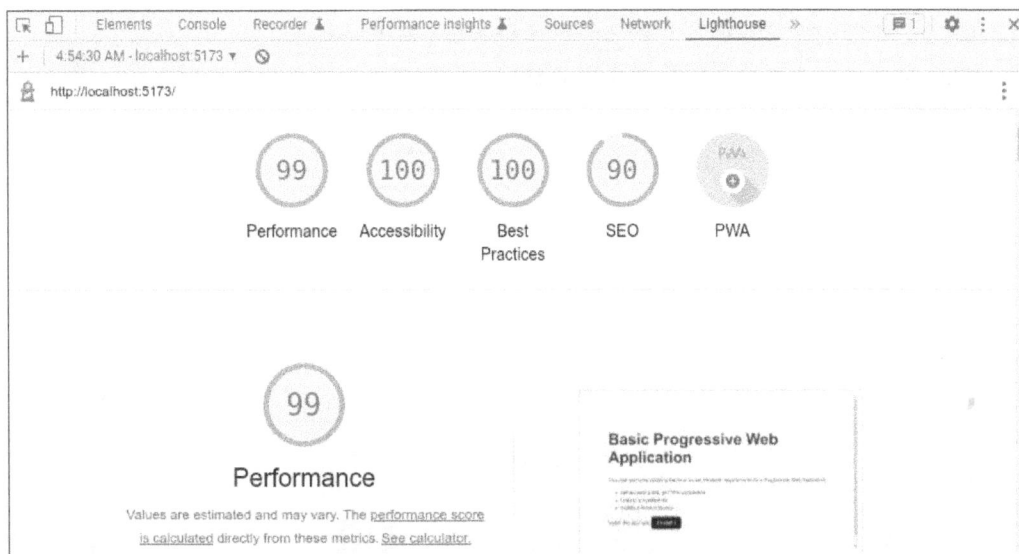

Figure 6.8 – Chapter 6 code example ratings in Lighthouse

Our example code application fully qualifies as a PWA and passes all tests with flying colors. This is easier to accomplish with smaller applications, of course. In practice, every rating above 90 is great.

Summary

In this chapter, we have taken a simple SPA and learned how to upgrade it into a PWA, manually and through the use of a plugin in Vite. Users can install PWAs in their platforms alongside native applications and interact with them, even if they are not connected to the internet. PWAs offer many advantages over web-only applications. We also saw how we can measure and rate our application in several industry-standard categories using Lighthouse. With this chapter, we end the incremental building of applications using web technologies and henceforth will focus on patterns and models for internal performance and efficiency.

Review questions

To help you solidify the concepts learned in this chapter, answer the following questions:

- What is the difference between a SPA and a PWA?

- What are the advantages of a PWA?

- What are the basic three requirements that a web application must comply with to be considered a PWA?

- Which tools we can use to incrementally prepare our application to be a PWA?

- What is a service worker, and what are some strategies to use it?

- What is a web manifest and why is it necessary?

Data Flow Management

In previous chapters, we have focused on understanding the Vue 3 framework and providing a context to create web applications. In this chapter, we will focus on the way our components communicate with each other and share information to make our application happen. We have touched on this topic briefly previously, but now we will dive deep into some patterns by implementing them alongside each other. Applying an appropriate information workflow is an important skill that can make or break an application. In particular, we will see the following approaches and code examples:

- Parent-child-sibling communication

- Implementing a message bus using the **Singleton** and **Observer** patterns

- Implementing a basic reactive state with composable components

- Implementing a centralized data repository with the powerful Pinia reactive store

- Reviewing browser-provided alternatives to share and store information

- Experimenting with reactivity, composables, and proxy patterns in action

As we have done previously, we will be building one concept at a time, incrementing in complexity. By the end of this chapter, you will have seen clear examples of implementation so that you can decide when to apply each one based on the needs of your application. Some of these are more suitable for small applications, and others for large, complex ones. You will be better prepared to control the workflow of information for your application.

Technical requirements

This chapter will approach concepts and apply patterns to control the communication and flow of information between components. You should be able to follow through with the code presented in this text, but for a better understanding and context experience, you would benefit from inspecting the full application code for this chapter, available in the repository for this book: `https://github.com/PacktPublishing/Vue.js-3-Design-Patterns-and-Best-Practices/tree/main/Chapter07`.

If you are starting a new project, just follow the instructions for scaffolding one, as seen in *Chapter 3, Setting Up a Working Project*.

Check out the following video to see the Code in Action: `https://packt.link/ZKTBJ`

Components' basic communication

We have seen previously that a parent component and its children have a rather simple and straightforward way to communicate. Parents pass data as `props` to their children, and these raise events (`emits`) to capture the attention of the parent. Much like the comparability of parameters and arguments in functions, `props` receive simple data by copy, and complex types (objects, arrays, and so on) by reference. We could pass, then, a plain object with member functions from the parent to the child, and have the child run the functions to access the parent's data. Even though this "works", it is sort of a dark pattern or anti-pattern, as it hides the relationship and makes it difficult to understand the data flow. The proper way to pass data upward in the component tree is through events (`emits`). Having said this, we must point out that child components are "ignorant" of each other, meaning that they do not have a direct way to communicate among themselves. We could pass a reactive variable and have each component involved access it, and this is certainly a working alternative, if not a clean one. In some cases, this would provide a simple solution, but again, it can lead to hidden side effects.

To manage in a clean way the workflow of data, we have several alternatives that follow good practices and design patterns. As a general rule and principle, the component that declares the variable is the owner of it, and it should be the one that manipulates it. With this in mind, in the most basic communication, the information needs to be maintained and manipulated by the parent component and shared among the children. We can leverage Vue's reactive system to spread the information. The key here is that only the parent component will manipulate it. Let's see how this works in practice with an example, implementing a small trivial application, as shown in *Figure 7.1*:

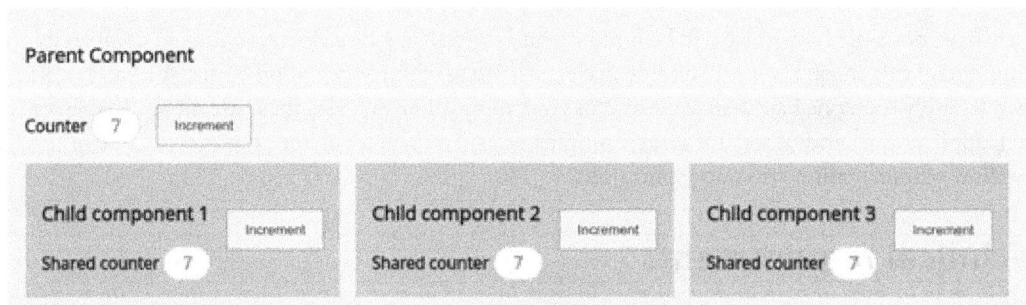

Figure 7.1 – Direct basic communication and reactivity

In this application, the parent component has three direct children and shares with them a reactive counter. All the components display a label with the value of the counter and have a button to trigger an increment... but only the father component performs the actual manipulation of the data. Vue handles the reactivity, meaning that when the parent modifies the value, the child components also receive them. Simple enough—let's see the important parts of how this is implemented:

/basic/ParentBasic.vue

```
<script setup>
import {ref} from "vue"
import ChildComponent from "./Child.vue"
const _counter = ref(0);                                    //1
function incrementCounter() {                               //2
  _counter.value++;
}
</script>
<template>
<div>
   <strong>Counter </strong>
   <span>{{ _counter }}</span>
   <button @click="incrementCounter()">                     //3
       Increment
   </button>
</div>
<section>
<ChildComponent title="Child component 1"
  :counter="_counter" @increment="incrementCounter()">      //4
</ChildComponent>
<ChildComponent title="Child component 2"
  :counter="_counter" @increment="incrementCounter()">
</ChildComponent>
<ChildComponent title="Child component 3"
  :counter="_counter"
  @increment="incrementCounter()"></ChildComponent>
</section>
</template>
```

In this component, we declare a _counter reactive variable (line //1) and an incrementCounter() function to manipulate its value (line //2). We trigger this function in the parent button, on the click event, as seen in line //3. Now, to see this pattern implemented, we just pass our reactive _counter variable as a prop to each child component, and we link our incrementCounter() function to each child's increment event (line //4). Simple enough—let's see how each child implements its part:

/basic/Child.vue

```
<script setup>
const
    $props=defineProps(['counter', 'title']),               //1
```

```
    $emit=defineEmits(['increment'])
function incrementCounter(){$emit("increment")}        //2
</script>
<template>
<h3>{{$props.title}}</h3>
<span class="badge">{{$props.counter}}</span>          //3
<button @click="incrementCounter()">                   //4
    Increment
</button>
</template>
```

Our child implementation is simple as well. We start by defining the props to receive the counter variable in line //1, and also our increment custom event so that we can notify the parent. In order to do that, we create a function in line //2. In our template, we display our prop in line //3 and trigger our increment function in line //4. Notice that our child component does not modify the counter. That is the responsibility of the father component, so we respect the pattern.

This pattern is one that we will use quite often, but it does have some limitations. For example, what happens when the data needs to reach a parent, sibling, or grandchild? Do we pass data up and down the tree, even though components don't use it? We could, but again, that is messy, verbose, and not the best way. We have better tools for that.

In *Chapter 4, User Interface Composition with Components*, we saw that a parent can pass data and functionality to any of their children down the tree, by using **dependency injection** (**DI**) with provide and inject. Since the example presented there was quite comprehensive, we will not repeat it here. I encourage you to review how the provision was created and injected. Instead of repeating ourselves, let's move ahead with the next item in our agenda to share information anywhere in the component tree: implement a **message bus** (also called an **event bus**).

Implementing an event bus with the Singleton and Observer patterns

A message bus is an implementation of the *Observer pattern* that we saw in *Chapter 2, Software Design Principles and Patterns*. As a short refreshment of the main concept, we seek to create an object or structure that receives and emits events that our components can subscribe and react to. This pattern runs independently of the component tree structure, so any *component and service* can make use of it. Visually, we can represent the resulting relationship as follows:

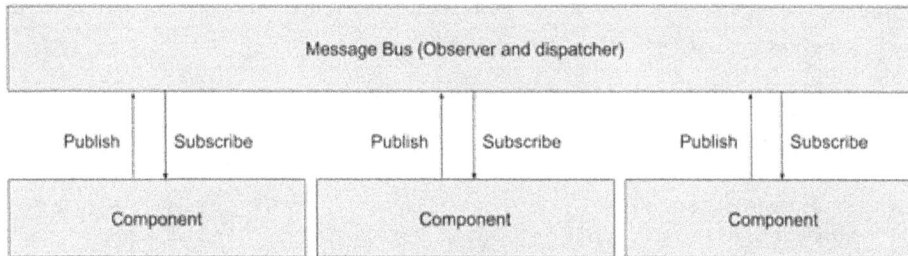

Figure 7.2 – A simplified view of a message bus relationship with components

From the preceding diagram, we can immediately see that each component is treated equally by the message bus. Each component subscribes one or more of its methods to a specific event, and at the same time has the same possibility to publish an event. This makes it very flexible, as events can also transport data.

Let's bring down to code these concepts with an implementation example. We start by creating a service, using the Singleton pattern, that provides us with a message bus. In our case, we will just wrap the mitt package, which gives us this functionality (see https://github.com/developit/mitt#usage).

The mitt package can be installed in our application with the following command in the terminal:

```
$ npm install mitt
```

Our service then looks like this:

/services/MessageBus.js

```
import mitt from "mitt"
const messageBus = mitt()
export default messageBus
```

This will give us a singleton for an event emitter and dispatcher, meaning our message bus. In our example, we will dispatch text messages through it, and each receiving component will display it. Our components will then look something like this:

/bus/Child.vue

```
<script setup>
import messageBus from '../services/MessageBus';          //1
import {ref, onMounted, onBeforeUnmount} from 'vue';
const
   $props=defineProps(['title']),
   message=ref("")                                         //2
```

```
    onMounted(()=>{
        messageBus.on("message", showMessage)})                          //3
    onBeforeUnmount(()=>{
        messageBus.off("message",showMessage)})
    function showMessage(value){                                         //4
        message.value=value;}
    function sendMessage(){                                              //5
        messageBus.emit("message",`Sent by ${$props.title}`)}
</script>
<template>
    <h4>{{$props.title}}</h4>
    <strong>Received: </strong>
    <div>{{message}}</div>
    <button @click="sendMessage()">Send message</button>                 //6
</template>
```

In this example, we start in line //1 by importing our messageBus object (check the right path in your implementation) and declare a message reactive variable initialized to an empty string. Notice how we also import and use the onMounted() and onBeforeUnmount() methods from the component's life cycle to subscribe and unsubscribe to the message event starting in line //3. The function that we register is in line //4, and it receives from the event a value that we pass to our internal variable to display in the template. We also need a function to publish the event to notify others, and that can be found in line //5. In this case, we publish the title of the component. This function is triggered by a button, as shown in line //6.

If you run the application example with some additional minimal styling, this code will result in something like this:

Message bus

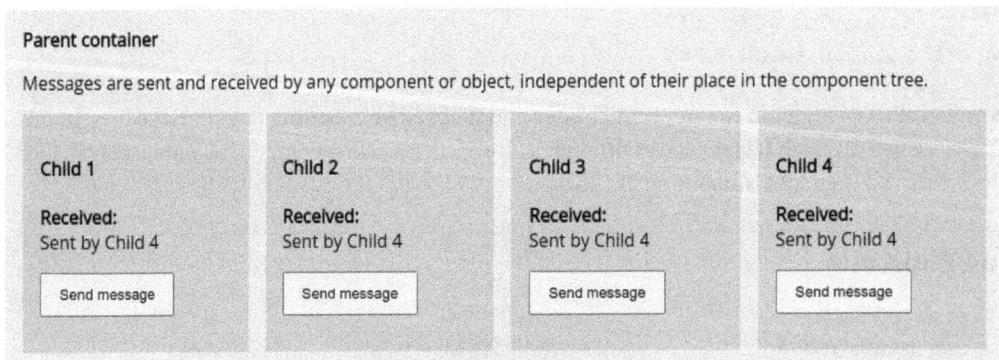

Figure 7.3 – A simple implementation of data sharing through the Observer pattern

This approach to handling the workflow of data is quite effective in what it does, but also has limitations. Events are a good way to notify multiple components simultaneously to trigger actions, independently of their place in the organization tree. When an application has multiple sub-systems that need to react to an application state change, this is a good pattern to apply. However, when dealing mainly with application data, this pattern has an important drawback: each component keeps an internal copy of the information. This makes the handling of memory quite inefficient, as the propagation of data means copying into different parts of our application. There are cases when this is necessary or desired, but certainly not for every case. If we have 50, 100, or 1,000 components subscribed to the same event, will all of them have the same copy of the data? If each component needs to handle and possibly modify the data independently of the others, this works fine... but if we want to make better use of Vue's reactivity and improve our memory handling, we need to use a different approach. This is what we will see next with a basic reactive application state.

Implementing a basic reactive state

As mentioned before, a drawback of using a message bus to share data is the multiplicity of copies of the same data, including the overhead for the handling of the events. Instead, we can leverage Vue's reactivity engine and, in particular, the `reactive()` helper constructor to create a single entity to hold our application state. Just like before, we can wrap this reactive object in a Singleton pattern to share it among components and plain JavaScript functions, objects, and classes. It's worth mentioning that this is one of the great advantages of Vue 3 and the new Composition API.

From the example code, we will end with a basic example like this:

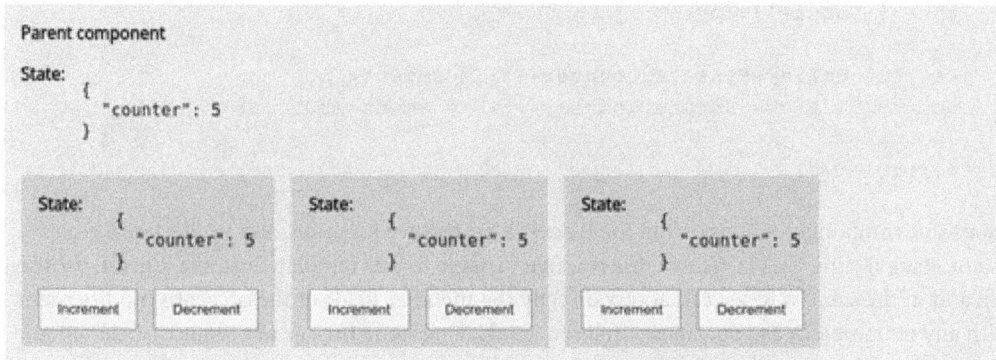

Figure 7.4 – A shared reactive object for state management

As you can see in the previous screenshot, the state in this case is shared (or accessed) by all the components of this example. Any of the child components can modify any of its values, and the change is reflected immediately across the application. In contrast with the previous examples, the implementation of this pattern is both simple and straightforward. Let's dive into it by first creating a service with our reactive state:

/service/SimpleState.js

```
import {reactive} from "vue"                          //1
const _state=reactive({counter: 0})                   //2
function useState(){return _state;}                   //3
export default useState;
```

If this code seems simple, it is because indeed it is. We create a JavaScript file and import the `reactive` constructor from Vue (line //1). Then, we declare a reactive constant with an initial object (line //2). This will be the application state that we return through a `useState()` function, named following the model of composable components (line //3). This function is our module exports.

Making use of this centralized state is also very simple, as we can see here:

/simple/ChildSimple.vue

```
<script setup>
    import useState from "../../services/SimpleState"    //1
    const $state=useState()
</script>
<template>
    <strong>State: </strong><br>
    <pre>{{$state}}</pre>                                 //2
    <div>
    <button @click="$state.counter++">Increment</button>  //3
    <button @click="$state.counter--">Decrement</button>
    </div>
</template>
```

We start our component by importing the `useState` factory function, and we declare a reactive constant using it (line //1). We use this reactive variable in our template just like any other (line //2), and in the same way, we can access directly the member fields of the object to modify them as with any other object, as you can see in line //3. Having done this, as one would expect, once a component modifies any value, the change gets propagated across the application.

This simple approach is very useful and fit for small to even medium-sized applications. It has many benefits, such as the following:

- It is easy to implement and understand.
- It leverages Vue's reactivity system.
- It is flexible, as we can add new reactive members after initialization.

- It establishes a single source of truth meaning that our state is the centralized repository of the application data. There's no need to keep internal or private variables synchronized.

If you consider the options we have seen until now, this is a giant leap forward. However, there are some situations when these falls short:

- What happens when a function modifies its value in an asynchronous manner if other components made changes before it was resolved?

- This approach does not allow us to handle computed data that needs to be implemented in each component

- Debugging could be hard as there is no specific support for developer tools

As mentioned, this approach is suitable for simple needs. For a more robust approach, we will dig into the official central state management solution by the Vue project: **Pinia**.

Implementing a powerful reactive store with Pinia

Central state management is not a concept private only to Vue, and the same pattern can be found in other libraries and frameworks. Just as in our basic reactive example, **Pinia** is a central state management tool that provides us with a single source of truth, meaning that a change in one of its values will propagate reactively to the entire application wherever is used. This state is shared among components in the application and gives us access to the full range of reactive tools Vue provides through a well-defined interface. It is easier to understand Pinia if first we build an example to show the results of using it. Running the code example will give us something like this:

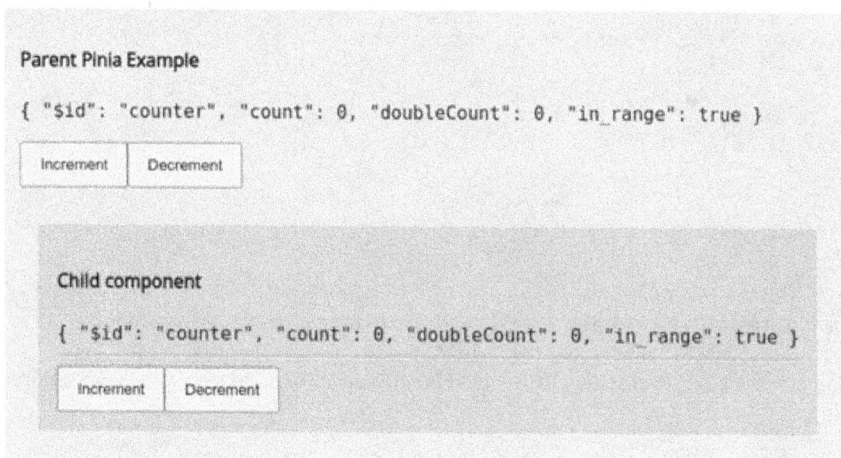

Figure 7.5 – Central state management with Pinia

In this example, we build a store that not only exposes a reactive state but also implements computed values. As an officially supported project, Pinia also exposes implementations of the Options and Composition APIs. To use Pinia, we need to first include it in our project with the following command in the project's root directory:

```
$ npm install pinia
```

After the installation, we should create a store and then attach it to our application so that it can be used by all the components. A store is like our reactive singleton from the previous section, meaning an object that will have reactive fields to be shared in our application, but also the related business logic. So, each store will have the following items: data, computed properties known as getters, and methods known as actions. We define each store in its own file as a module, defining each item. Using the Options API, a store would look like this:

Options API basic store

```
import { defineStore } from 'pinia';                         //1
const useCounterStore = defineStore('counter', {             //2
  state: () => {return {count: 0, in_range: false}},         //3
  getters: {
    doubleCount: (state) => {                                //4
      if(state.count>=0){
            return state.count *2;
      }else{

       return 0
      }
  }, inRange: (state)=>return state.count>=0},
  actions: {                                                 //5
    increment(){this.count++},
    decrement(){this.count--;}
  },
})
export {useCounterStore}
```

In this store, we start by importing the defineStore constructor from the Pinia package (line //1) and use it to create a store in line //2. This constructor receives two arguments:

- The name of the store, as a string. This has to be unique among the stores, as it is used internally as an ID.

- An object with the store definition with the following members:

  - state (line //3): This is a function that returns an object. Notice that we do not declare it to be reactive. Pinia will take care of that.

- `getters` (line //4): This is an object whose members will become computed properties. Each member receives as the first argument the state of the store, as a reactive object.

- `actions` (line //5): This is, again, an object whose members are functions that can access and modify the state but must do so by accessing it through the `this` keyword.

Using the Options API to define the store is a good way to understand the parts that make it up. However, the change of syntax between `getters` and `actions` could be confusing and lead to involuntary mistakes, as one accesses the state through an argument and the other by using the `this` reference. However, if we take a moment to look at the constructor, we can see that `getters` and `actions` are analogous to *computed properties and component methods* (functions). With that in mind, let's see how to rewrite this store using the Composition API, and this is the one we will use in our example code:

/stores/counter.js

```
//Composition API
import {ref,computed} from 'vue'                           //1
import {defineStore} from 'pinia'

const useCounterStore=defineStore('counter',()=>{          //2
    const
        count = ref(0),                                    //3
        in_range=ref(true),
        doubleCount = computed(() => {                     //4
            if(count.value>=0){
                return count.value *2;
            }else{
                return 0
        }}),
        inRange = computed(()=>return count.value>=0);
    function increment() {count.value++}                   //5
    function decrement(){count.value--;}
    return {                                               //6
        count, doubleCount, inRange,
        increment, decrement
    }
})
```

Using the Composition API makes the store look more like the rest of our application, as we apply the same approach. We start by importing from Vue the constructors we need in line //1, as with components using the same API. This time, when we use the `defineStore` constructor, instead of passing an object we pass a function (or arrow function) that will return the reactive properties and methods that make up the store. You can see this in line //2, and then the `return` object in line //6. As you can expect, inside that function we declare our reactive properties (line //3) and computed properties (line //4), and methods (line //5). Reactive properties will become, well, reactive properties. Computed properties will become our getters, and the functions will become the actions. This far, this syntax does not have the syntactic sugar we are used to using the `<script setup>` tag, but the body of the function is the same approach (state of mind) that we use with components.

Now that we have a store (and we could have many), before we can actually use it, we need to implement Pinia in our application. For that, in our `main.js` file, include the following highlighted lines:

./main.js

```
import { createApp } from 'vue'
import { createPinia } from 'pinia'
import App from './App.vue'
const app = createApp(App)
app.use(createPinia())
app.mount('#app')
```

This step is necessary to enable the Pinia engine for the entire application. What is left now is to import the store we want to use in our components that need it. For example, if you look into the example repository, you will find this file:

/pinia/ChildPinia.vue

```
<script setup>
import { useCounterStore } from '../../stores/counter';    //1
const $store=useCounterStore()                             //2
</script>
<template>
    <h4>Child component</h4>
    <code :class="{'red': !$store.in_range}">              //3
        {{$store}}
    </code>
    <button @click="$store.increment()">                   //4
        Increment</button>
    <button @click="$store.decrement()"
        :disabled="!$store.in_range">Decrement
    </button>
</template>
<style scoped>
.red{color: red;}
</style>
```

We import the store constructor in line //1, and we create our reactive object in line //2. To use their values or execute their methods, we use them directly as if they were regular objects using the dot (.) notation. Notice in line //3 how we access the value of in_range, and later, in line //4, we execute the `increment()` function. As we would expect, any modification of the store values will be synchronized automatically across our application.

Unlike previous methods, Pinia stores and states are traceable and show up on the developer tools. For applications of medium size and above, using Pinia is almost a requirement when a central state is necessary.

Pinia is Vue 3's official solution for central state management, replacing Vuex from the Vue 2 branch. In practice, they accomplish the same functionality, but the former has some advantages that made the Vue team select it and sponsor it. A deep review is not a topic for our purposes, but here is a short list of changes or advantages of Pinia:

- Different approach to stores. In Pinia, each store is its own module, and they are all dynamic. Vuex instead had one single store, with partitions in modules.

- The syntax and API for Pinia are simpler and less verbose than Vuex.

- Better support for TypeScript and discoverability for the IDE's autocomplete features.

- Support for both Options and Composition APIs.

- Better internal use of Vue's new reactive models.

- Developer tools support.

- A plugin architecture to extend Pinia.

The change from Vuex to Pinia makes it difficult to make a one-step replacement upgrade for projects that were using it. However, the Pinia team has published a nice migration guide on the official website that you can find here: `https://pinia.vuejs.org/cookbook/migration-vuex.html`. For a complete reference of all the options available with Pinia, I recommend reading the official documentation at `https://pinia.vuejs.org`.

With Pinia, we have seen now the most common and relevant patterns to control the flow of data between components (and services!), but these are not the only ones available to us. We will see next the stores provided by default in modern web browsers, and how to use them.

Browser data stores – session, local, and IndexedDB

Browsers provide other features to store data locally, which can be read not only by any other component but also by any script running on the same page. We will not talk about cookies, but the new methods provided as key-value stores: `SessionStore` and `LocalStore`. But these are not the only options, as browsers also provide a database called `IndexedDB` that offers much more storage space and can be accessed also outside the scope of our application's window in a different thread. We will see how in *Chapter 8, Multithreading with Web Workers*, in more detail, while here, we will focus first on understanding the basic concept and limitations of each one.

`SessionStorage` is a read-only object created for each page origin. It stores only string data that can be accessed and retrieved using a simple interface. This data exists only through the duration of the *browser tab* and persists during refreshes. A clear example of this use is to persist form data. The object is attached to the `window` object (`window.sessionStorage`) and can be accessed by any script on the page.

`LocalStorage` is similar to `SessionStorage` in capabilities and data storage. It has the same interface and is restricted also to the same origin of the page. The main difference is that it persists beyond the life of the page and is shared among all the open pages of the same origin. Websites and applications can use it to store data and retrieve it throughout multiple sessions on the same browser.

`SessionStorage` and `LocalStorage` share the same interface:

- `.setItem(item_name, item_data)`: Here, `item_name` is a string that uniquely identifies `item_data`, which is also a string
- `.getItem(item_name)`: Retrieves the string data stored under `item-_name`, or null if not found
- `.removeItem(item_name)`: Deletes the data by `item_name` from the store
- `.clear()`: Removes all data from the store

The preceding methods represent the totality of API endpoints for both storages. Simple enough—we can serialize data to record it in these stores. For example, to store a JSON object, we would use the following (we can omit the `window` object reference, as it is considered a global object):

```
localStorage.setItem("MyData", JSON.stringify({…}));
```

And then, to retrieve it, we would use the following:

```
let data=localStorage.getItem("MyData")
if(data){
    data=JSON.parse(data);
}
```

Both stores have some limitations and a few caveats:

- There is no standard limit set among browsers for how many characters each store can hold. Strings are stored in UTF-16, so each character can take from 2 bytes or more (see `https://en.wikipedia.org/wiki/UTF-16`), which makes calculation hard. The specifications recommend at least 5 MB for each storage.

- When these storages run out of space, some browsers crash the page, while others prompt the user for consent to expand the storage.

- Access to store and retrieve data is sequential, possibly blocking the render process and making the page/application look irresponsive... But this only happens in long operations.

- For `sessionStorage`, duplicating tabs will also duplicate the storage. Instead, for `localStorage`, both tabs will access the same information.

- Neither localStorage nor sessionStorage is reactive or provides listeners to watch when a value changes.

The preceding limitations are in no way a threat or a suggestion not to use them. Instead, they are the boundaries and limits to using them, since all data is stored locally on the user's browser, and nothing is sent back to the server (as cookies do).

In contrast to these web storage objects, `IndexedDB` is a different system altogether. It is a full implementation of a transactional database that stores JavaScript objects under a unique key. We can open multiple databases, create connections to them, and define schemas, and all operations are asynchronous, so there is no application blocking. The size limit has also been extended, with a soft limit of 50 MB. If a database grows more than that, the user is prompted to consent to expand it, and more space is given. In theory, depending on the implementation in each browser, it could occupy as much space as available. In practice, each browser has its own way to negotiate available space with the local operating system, so no hard number can be given about its limits that would hold true in every case.

> **Curiosity**
>
> The Chrome engine provides a flag to build the engine without limits to `IndexedDB`, save for the available disk space. This flag can also be activated in hybrid frameworks such as NW.js or when building the browser from source.

There is a major issue with `IndexedDB`, which is that its API is complicated and cumbersome, so it is very rare that an application would access it directly. Instead, since `IndexedDB` is so flexible and fast, there is a fair number of libraries that create their own database implementation on top of it or facilitate a simpler interface (using the Façade pattern, for example). A curated list of these libraries and frameworks can be found in the **Mozilla Developer Network** documentation (`https://developer.mozilla.org/en-US/docs/Web/API/IndexedDB_API#see_also`). In our implementation examples for *Chapter 8, Multithreading with Web Workers*, we will use one of these libraries. For the purposes of this chapter, just keep in mind that each browser provides you with a powerful database for your application, and you can access it through a variety of patterns and approaches.

Experimenting with reactivity and Proxies patterns

It is time to put into practice what we have learned in this chapter under the light of patterns we saw in *Chapter 2, Software Design Principles and Patterns*, with a small experimental project. We want to create an option to make `sessionStorage` data behave like a reactive central state manager so that we can use it in our components. Possible uses for this approach could be to persist user-entered data during refreshes, alert components of data changes, and so on.

Since `SessionStorage` does not provide an API we can listen to, our approach will be to create a Proxy handler using the Decorator pattern, to match and keep synchronized the values in the store with an internal and private reactive property. We will wrap this in a *singleton* and use the *Central State* manager approach to share it in our application. Let's start by creating our core service:

/services/sessionStorage.js

```
import { reactive } from 'vue';
let handler = {                                              //1
    props: reactive({}),                                    //2
    get(target, prop, receiver) {                           //3
        let value = target[prop]
        if (value instanceof Function) {
            return (...args) => {
                return target[prop](...args)
            }
        } else {
            value = target.getItem(prop)
            if (value) {
                this.props[prop] = value;
            }
            return this.props[prop]
        }
    },
    set(target, prop, value) {                              //4
        target.setItem(prop, value)
        this.props[prop] = value
        return true;
    }
}
const Decorator= new Proxy(window.sessionStorage, handler);  //5
function useSessionStorage(){                                //6
    return Decorator;
}
export { useSessionStorage }
```

In this `service` module, we will use the native JavaScript implementation of a `Proxy` object to capture specific calls to the API of the `window.sessionStorage` object. The use of Proxy objects is rather advanced in JavaScript, so I recommend that you look at the documentation on MDN here:

`https://developer.mozilla.org/en-US/docs/Web/JavaScript/Reference/` `Global_Objects/Proxy`. We start by importing the `reactive()` constructor from Vue and then creating a plain object named `handler` (line `//1`), that will act as our proxy/decorator. This object will be placed to intercept the calls to the original `sessionStorage`. Inside it, we declare a `prop` property as reactive (line `//2`)), initializing it with an empty object. This object will be synchronized with the storage. Then, we create two traps (or interceptors): one for get or read operations (line `//3`), and another for set or write operations (line `//4`).

The `get()` function receives three arguments, of which we will use only two. The target refers to `sessionStorage`, and `prop` is the name of the method or attribute requested. Because `prop` can be either one, we test if it is a function with an `if` statement, and if so, we return a function that takes in all the arguments and returns the original function call with them. If it is not a function, then we retrieve the item from the store, test if it is part of our internal reactive property, and finally, return the value. This makes sure that our internal `props` object is in sync with values created before the decorator was implemented.

The `set()` function is simpler, as we just take the value passed and store it in both places: our internal props and the store.

With our handler ready, in line `//5`, we create a `Decorator` proxy object using a native JavaScript constructor and provide a `useSessionStorage()` function in line `//6` so that we can export it as a singleton.

With our Decorator created, now we can use it in our components, with the same approach as is standard in Vue 3:

/session_storage/ChildSession.vue

```
<script setup>
    import {useSessionStorage} from "../../services/SessionStorage"
    const $sessionStorage = useSessionStorage()
</script>
<template>
    <strong>Child Component</strong>
    Counter: {{ $sessionStorage.counter }}
</template>
```

Notice that now we can use this object as a Pinia store or a simple reactive object, and the value of `sessionStorage` will always be synchronized and persist even if we refresh the page. To view the full example, please check the implementation of the code example in the GitHub repository. When you run it, you will see a section like this one:

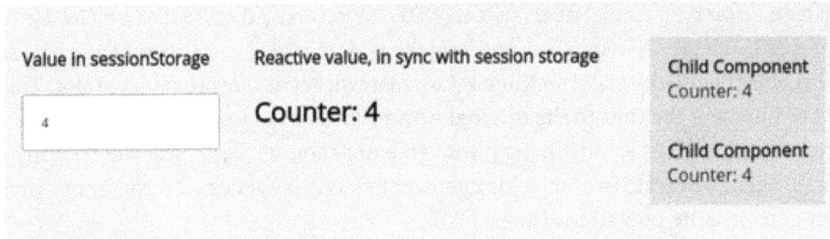

Figure 7.6 – Example of our reactive $sessionStorage object

In this example, we also implemented a parent component with an input element. When you modify the value, it is automatically synchronized and reflected in the children component, but also in sessionStorage. If you open the browser's developer tools and navigate to the **Web Storage** section, you will see this reflected. Here is a screenshot of how this looks in Chrome, on an Ubuntu system:

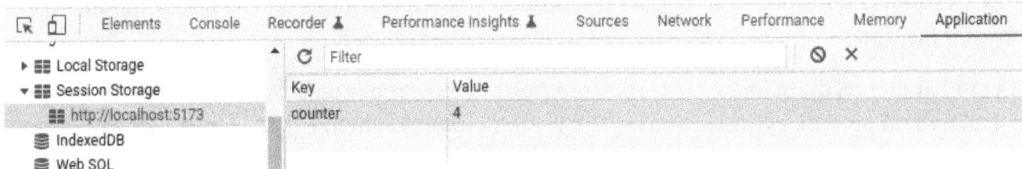

Figure 7.7 – Session Storage showing the item from the example

In the same way we implemented this pattern for the session storage, we could also, with a few changes, apply it to local storage.

Summary

In this chapter, we have seen in detail the different approaches and methods to control the flow of data between our components, services, and persistent storage provided by modern browsers. We also took time to integrate our knowledge by experimenting with session storage and the Decorator pattern to create a reactive/persistent central state. We took time to differentiate between approaches, and we have seen code for the implementation of each method. All these new skills are used daily in the development of Vue 3 applications.

In the next chapter, we will investigate improving the performance of our application by using advanced JavaScript tools: web workers.

Review questions

Use these questions to review what you have learned in this chapter:

- Which methods do we have available to share data between sibling components?

- What is a message/event bus, and when is it most useful?

- What is a central state management approach, and how can we implement it?

- What is the difference between session and local storage?

- How can we see what information is stored in session or local storage?

8

Multithreading with Web Workers

In this chapter, we will cover important topics that will highly improve the performance of a web application, especially single-page applications. First, we will learn how websites and JavaScript work, and how to use web workers to leverage our application processing power, data access, and network communications. Then, we will introduce two new conceptual design patterns and will implement them in an example application together with other patterns that we have previously seen. On top of this, we will also introduce two libraries that will facilitate our network communications as well as the handling of our persistent database(s) in IndexedDB. We will also implement a simple Node.js server to provide us with feedback and test our work in a highly decoupled architecture, where our frontend and backend services communicate using standard APIs over the HTTP protocol.

In this chapter, we will cover the following topics:

- WebWorkers

- Business and dispatcher patterns

- Network communication inside a `WebWorker`

- A browser's persistent embedded database – IndexedDB

- How to build a simple Node.js API server for testing

The concepts in this chapter can be considered "advanced," but we will condense them down into understandable pieces that we will implement right away. By the end of this chapter, you will have a solid knowledge of how to implement multithreading in your web applications and also a reference framework to scale and facilitate the use of complex browser APIs.

Technical requirements

This chapter does not add additional requirements to our application. However, we will only see relevant parts of the code, so to see the entire application working, you should refer to the code examples for *Chapter 8, Multithreading with Web Workers*, in the book's GitHub repository at `https://github.com/PacktPublishing/Vue.js-3-Design-Patterns-and-Best-Practices/tree/main/Chapter08`.

Check out the following video to see the Code in Action: `https://packt.link/D4EHt`

An introduction to web workers

JavaScript is a single-threaded language, meaning that it doesn't natively have a way to spawn processes in separate threads. This makes web browsers run the JavaScript in a web page on the same thread as other processes, which directly affects the performance of the page, most notably, the rendering process,that is in charge of presenting the page on the screen. Bowsers make a considerable effort to optimize the performance of all these moving parts to make a page responsive, performant, fast, and efficient. However, there are tasks that a web application must do in JavaScript that are heavy and potentially "render-blocking". This means that the browser will have to pay attention to the results of the code and use all the resources to complete the running function before it can focus on the rendering (presenting the page to the screen). If you ever find a process on a web page that makes the site seem "unresponsive" or "stuttering" after you start an action (your mouse may even freeze in some cases), this could be one of the causes.

If we open the developer tools in a modern browser, we can access some performance tools to analyze how a web page behaves and how much time each process step takes. For example, here is a quick view of the first load of YouTube on a shared link, in Firefox for Linux:

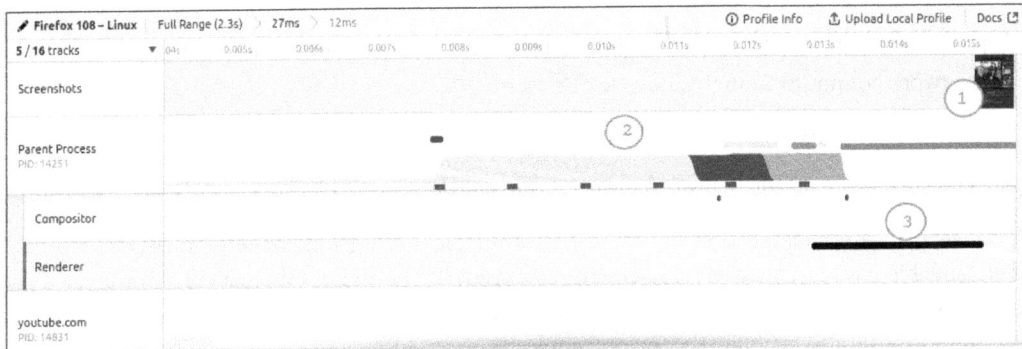

Figure 8.1 – The performance of YouTube's first load, seen using developer tools

The preceding screenshot has zoomed into the actual processing of the page, showing what happens before the first rendering, meaning before the user can actually see something on the screen. That is represented in the first line, **Screenshots**, where for this case, the first visible elements appear towards the end of the timeline (*#1*). The second line shows what the main **Parent Process** has been busy doing, and if you pay attention, the very first section (*#2*) is all about processing JavaScript. The **Renderer** process, highlighted and displayed with a black bar (#3), can't even start until the JavaScript has been run. When it does run, it draws the page on the screen, and you have the visible content from *#1*. This gives an approximate idea of the work the browser does each cycle in between screen paints (called "frames"). The browser attempts to produce as many **frames per second (fps)** as possible. To keep a fluent 60 fps, it needs to do all this processing in about 16.67 milliseconds or less. At best, your JavaScript process should be resolved in half that time to keep a fluid experience for the user. With this in mind, what happens when your JavaScript takes longer than that? Simple enough, the render process is postponed, the fps drop, and the user experiences stuttering and a frozen **user interface (UI)** may happen. Depending on your web application, this could be an important issue.

You may say, "Wait a minute, why we don't make heavy tasks asynchronous? Wouldn't that resolve the issue?" The answer is: maybe and no. When you declare an asynchronous function, it only means that the execution will be deferred to a place in time when the processing of the sequential code has been executed. Most likely, this pushes the asynchronous code toward the end or after the sequential code has been executed, but then it will be run sequentially as usual. If the rendering process happens before that, you may perceive a performance gain, but if not, you face the same issue if the async function takes longer (as it will affect the next rendering). If we moved all functions to be asynchronous, we would end up potentially with the same result as if everything were sequential, plus the overhead of making the asynchronous calls:

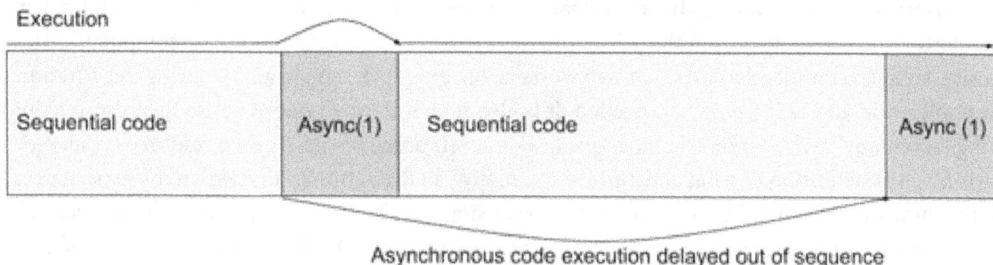

Figure 8.2 – A representation of the execution of async code, moved
after the sequential code has been executed (1)

Then, if asynchronous operations would not completely solve the performance issue, how do we resolve it? Beyond all the optimizations you could make, there is one technology you should also consider at the top of the list of alternatives: the web workers API.

Web workers are JavaScript scripts that execute in their own process (or thread, depending on the implementation); thus, they do not compromise the parent process where the rendering happens. The browser API provides a rather simple yet effective way to communicate to and from the parent process: a messaging system. These messages can only pass serializable data. The parent process and each web worker operate in their own environment and memory boundaries, so they cannot share references or functions, hence why all the data passed between them has to be serializable as it is copied into each process. While this may seem like a disadvantage, it is actually an asset when used properly, as we will see soon here. Another caveat of this architecture is that web workers do not have access to the **Document Object Model** (**DOM**) or the Window objects and, consequently, to any of their services. They do, however, have access to the network and IndexedDB. This opens up a wealth of opportunities for the architectural design of your frontend application, as you can easily separate what a presentation layer and a business layer is.

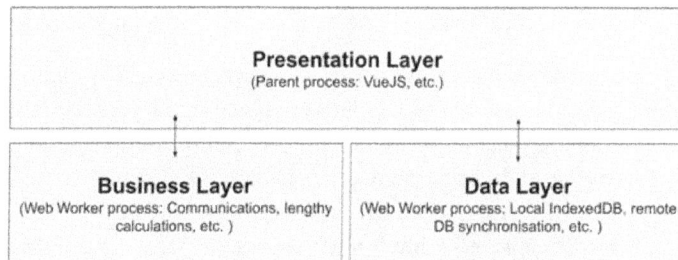

Figure 8.3 – A layered representation of a Vue application with background processes using web workers

As you can see in the previous diagram, we can instantiate multiple web workers to represent different types of layers in our application (**Business**, **Data**, **Communication**, and so on). While a web worker can be started and terminated from the parent process at will, both of these actions are computationally expensive, so the recommendation is that web workers, once created, remain active during the duration of the application and accessed when needed. It is also recommended not to abuse this resource by creating "too many" web workers, as each one is a different process with its own resources reserved. There is no clear definition of what constitutes "too many", so discretion is advised. In my experience, while the number of web workers remains in the lower single digits, even low-powered devices should handle your application with excellent performance. As with many other things, there can be too much of a good thing, and this also applies with web workers.

Now that we know what web workers are and what they can do for us, let's see how to implement them in pure JavaScript and then how to do so with Vite.

Implementing a Web Worker

Creating a web worker in plain JavaScript is quite simple and straightforward. The window object provides a constructor, properly named Worker, which receives as a parameter the path to a script file. For example, considering that our web worker is contained in a my_worker.js file, this is how we can create it:

```
if(window.Worker){
    let my_worker=new Worker("my_worker.js")
    ...
}
```

Simple enough, if the constructor exists in the `window` object, then we just create a new worker accessing the constructor directly. The newly created worker again exposes a simple API:

- `.postMessage(message)`: This will send the message to the web worker. It can be any data type that can be serialized (basic data types, arrays, objects, and so on).

- `.onmessage(callback(event))`: This event is triggered when the workers send a message to the parent process. The event received has a `.data` field that contains the message/data passed by the worker.

- `.onerror(callback(event))`: When an error occurs in the worker, this event is triggered, and it will contain the following fields:

  - `.filename`: With the name of the script filename that generated the error.

  - `.lineno`: The line number where the error occurred.

  - `.message`: A string containing the description of the error.

This messaging system allows us to carry on what otherwise could be a very complex form of **inter-process communication (IPC)**. Our previous code should look as follows due to implementing it:

```
let my_worker=new Worker("my_worker.js")
my_worker.onmessage=event=>{
    // process message here
    console.log(event.data)
}
my_worker.onerror=err=>{
    //process error here
}
my_worker.postMessage("Hello from parent process");
```

To complete this, we now need to implement the `my_worker.js` script. For this example, it can be something as simple as this:

./my_worker.js

```
self.onmesssage=event=>{
    console.log(event.data)
})
setTimeout(()=>{
    self.postMessage("Hello from the worker")
},3000)
```

Our example worker is very simple. It prints the data received to the console, and 3 seconds after it has been activated, it sends a message to the parent process. Notice that we are using the `self` reserved word. This is needed when accessing the API from within a function, as it references the worker itself. This is why it is necessary inside the `setTimeout` callback. At the root level, it is optional, so you can write `self.onmessage` as in our example or directly `onmessage`.

Web Workers can instantiate other workers and also import other scripts through the `self.importScript()` method or just `importScript()`. This method receives a string with the script filename as a parameter. This is analogous to how we use the `import` statement in our services and components in our main application.

When using Vite, as we are doing to bundle our Vue application, we have an alternative way to import and create a worker by using a suffix. For example, add the following in our `main.js` script:

./main.js

```
import MyWorker from "my_worker.js?worker"
const _myWorker=new MyWorker()
_myWorker.postMessage("Hi there!")
_myWorker.onmessage=(event)=>{...}
```

When using the `worker` suffix notation, Vite wraps the implementation in a constructor, which we can use to instantiate our worker. This way makes handling workers more akin to using any other class in our application, as we can use the same approach to include it in our application, and this is the syntax we will use in our examples. Additionally, Vite will process the scripts from our worker, so we can use our more familiar syntax to import resources (`import ... from ...`) instead of the native `self.importScript()`.

There is more about web workers to learn. For our purposes, this is enough and what we will use. If you would like to know more, please refer to the documentation on the Mozilla Developer Network (`https://developer.mozilla.org/en-US/docs/Web/API/Web_Workers_API/Using_web_workers`).

With these building blocks, we can now implement a robust and easier-to-handle connection to our web workers by applying design patterns. Before we do that, we need to learn two more patterns conceptually: the business delegate and the dispatcher patterns.

The business delegate pattern

This pattern is used to hide the complexity of accessing business services or a business layer from the client or presentation layer by providing a single point of access with a well-defined and simple(r) interface. It can be reasoned to some degree as a variant or evolution of the proxies and decorator patterns that we saw in *Chapter 2, Software Design Principles and Patterns*, but applied at a larger logical scale between architectural layers. It usually involves at least the following entities:

- A **business delegate** entity, which acts as the single point of entry for the client to all the available services

- A **business lookup or router** entity whose function is to route the execution of the incoming request to the appropriate service

- The **business services** that expose a common interface (directly or via a proxy pattern) with the provided function

The pattern can be represented for our purposes in the following diagram:

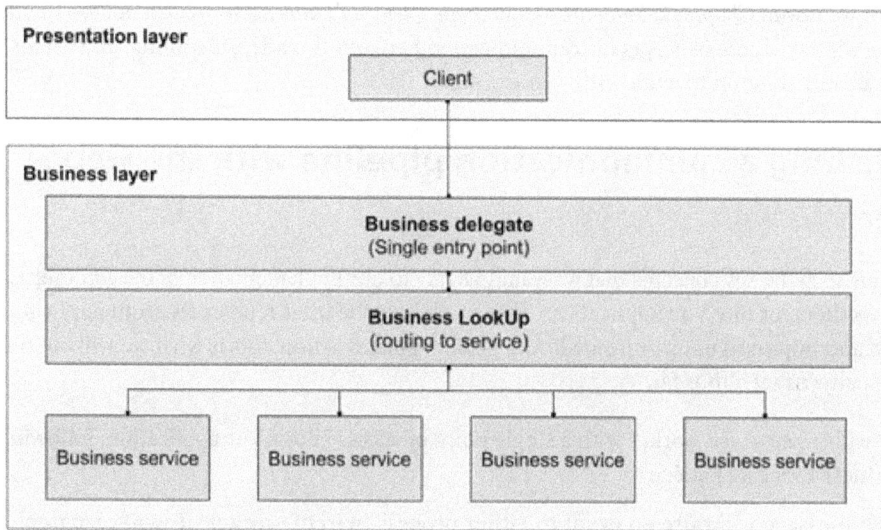

Figure 8.4 – A representation of the business delegate pattern

This pattern can be applied to multiple architectural levels. In our particular case, we want to apply this design to our application with web workers. We will consider the parent process as our presentation layer and our web worker as our business layer. In the parent (or main process), we will have our Vue application, as usual, focusing primarily on providing an excellent user experience. The worker will then be responsible for providing us with access to services, either local, as in the case of the IndexedDB, or remote, encapsulating the communication with our server and additional services and any additional computationally intensive function. This separation of concerns has many advantages, not only from the performance point of view but also from the design and implementation of the application as a whole.

Before we implement the code for this chapter, we need to see one more pattern that we will implement since we can only pass serializable data between processes and cannot execute function calls as a plain Business Delegate pattern proposes. We will expand on the idea of the command pattern and use what is called a dispatcher pattern.

The dispatcher pattern

We have seen previously that either our parent process or the web worker process can initiate communication by posting (sending) a message to each other. As long as the proper listener (`onmessage`) has been defined, either one can receive and react to these events. In the Dispatcher pattern, these messages contain information related to the event, such as data. The key factor that separates this design pattern is that *the event messages must be published between threads and scheduled for execution upon arrival*. Of course, this scheduling can also include the "immediate execution" of some task or function.

The implementation of this pattern is rather trivial, and you may think of it as akin to the Command Pattern that we saw in *Chapter 2, Software Design Principles and Patterns*, so we will not see this again. Instead, we will take these concepts of communication between threads, scheduling, and events with data to create our solution to work with web workers.

Establishing a communication pipeline with the web worker

We have now seen the key concepts that we want to apply to our implementation of the communication with web workers for our Vue application. This model can be used repeatedly from application to application and improved upon as needed. As a general plan of action, this is what we will build using the design patterns seen thus far:

- We will create a web worker with a single point of access in our Vue application, following the Business Delegate pattern
- Each message will raise an event to either process (parent-worker or worker-parent) and include command and payload data, as well as tracking information for scheduling as in the Dispatcher Pattern

Simple enough, the architecture described in the preceding points allows us to establish a workflow as shown here:

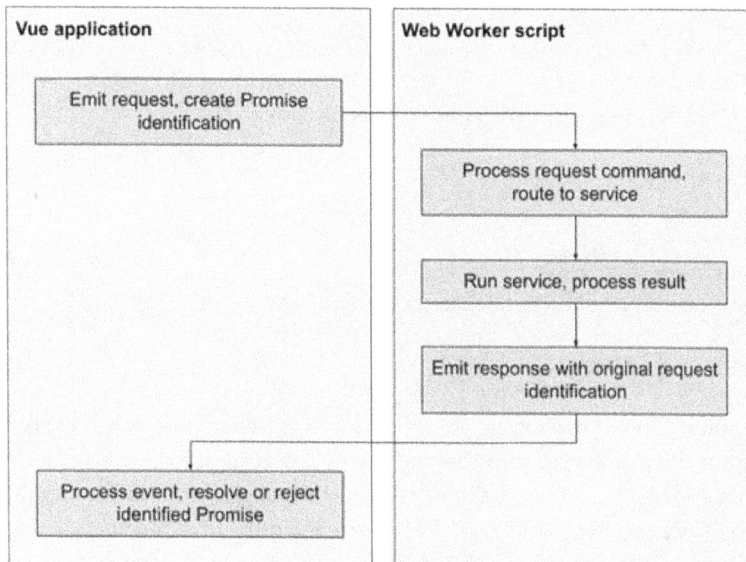

Figure 8.5 – Implementation of the communication workflow with the web worker

Now that we have the theoretical basis and a broad understanding of what we will create, it is time to move into the code. We will focus on the most relevant parts of the code that implement the model mentioned previously. To see the entire application code, please review the full source code from the GitHub repository. Let's start by creating a service that will be our entry point in the client application:

./services/WebWorker.js

```
import WebWorker from "../webworker/index.js?worker"
const _worker = new WebWorker()                                    //1
const service = {
    queue:{},                                                      //2
    request(command, payload = {}) {                               //3
        return new Promise((resolve, reject) => {                  //4
        let message = {
            id: crypto.randomUUID(),
            command,
            payload
        }
        service.queue[message.id]={resolve, reject}                //5
        _worker.postMessage(message);                              //6
        })
    },
    processMessage(data) {
        let id=data.id
```

```
            if(data.success){
                service.queue[id].resolve(data.payload)          //7
            }else{
                service.queue[id].reject(data.payload)
            }
            delete service.queue[id];                             //8
        }
    }
    _worker.onmessage = (event) => {
        service.processMessage(event.data);                       //9
    }
    export default service;                                       //10
```

This implementation is simple yet effective. It serves us well to understand how these patterns work. We start by importing the web worker constructor using Vite's special suffix worker and then creating the instance reference in line //1. As usual, this service will be a singleton, so we create it as a JavaScript object, which we will export later in line //10. The service has only three members:

- queue: This is defined on line //2 and is a dictionary that we will use to store our scheduled calls to the web worker using a unique identification. Each entry will save the reference to the resolution methods of a promise (resolve and reject).

- The request() method: Defined on line //3 here, this will be used by other services and components (the "clients") to request tasks from the web worker. It always returns a promise (line //4). The message passed to the web worker encapsulates the command and payload received as parameters with a unique identification. We save the reference to the resolve() and reject() methods in the queue (line //5), and finally, using the native messaging method of the web worker, we post the message on line //6.

- The processMessage() method: This receives the data submitted by the web worker, and based on the identification and the result of the operation passed in the .success attribute (Boolean), we access queue and either use the resolve() or the reject() function to resolve or reject the promise (line //7). Finally, we remove the reference from the queue in line //8.

The last step in this file is to link the incoming messages passing the data directly from the worker to service.processMessage() in line //9. It may be clear by now that we have made some decisions regarding the structure of the message and also the reply. Messages have three components: id, command, and payload. Replies also have three elements: id, success, and payload. On the client side, we have chosen to operate with promises, as they do not "time out".

With the client side resolved, now it's time to work on the web worker script. Create the following index.js file in the webworker directory:

./webworker/index.js

```
import testService from "./services/test"
const services= [testService]                              //1
function sendRequest(id, success=false, payload={}){
    self.postMessage({id, success, payload})               //2
}
self.onmessage= (event)=>{                                 //3
    const data=event.data;
    services.forEach(service=>{                            //4
        if(service[data.command]){                         //5
        service[data.command](data.payload)                //6
            .then(result=>{
                sendRequest(data.id, true, result)         //7
            }, err=>{
                sendRequest(data.id, false, err)
            })
        }
    })
}
```

The web worker is even shorter, and we have also made some decisions regarding the interface implemented by each underlying service: each of their methods has to return a Promise as well. Let's see the code and find out why.

We start on line //1 by importing testService (we will create it later) and include it in an array of services. This will make it easier to add new services by importing them and just including them in this array (this could be a stepping stone to a plugin architecture but we'll stay simple for now). We then define a sendRequest() global function, which will send a message to the parent process with a coded message with three fields: id, success, and payload, as expected by the client in our defined . This is what happens in line //2.

In line //3, we define the onmessage event handler to process the incoming messages. When one is received, we traverse our services array to find a matching command (line //4), and when we do (line //5), we execute the function by passing the payload as a parameter (line //6) after we parse it through the JSON utility. Then, with the resolution or rejection of the promise, we transmit the proper result to the client in line //7. This short piece of code acts as the *business delegator and dispatcher*. Finally, let's take a look at testService to see how it works:

./webworker/services/test.js

```
const service={
    test(){
        return new Promise((resolve, reject)=>{
            setTimeout((()=>{
                resolve("Worker alive and working!")
            }, 3000)
        })
    }
}
export default service;
```

As you can appreciate, this test service doesn't do much other than just return a Promise and set a timer to resolve it after 3 seconds. This delay is artificial since, otherwise, the reply would be immediate. If you run the example application, when you click the **Send request** button, you will see the message changing from **Waiting...** to **Worker alive and working!** after 3 seconds, just as expected:

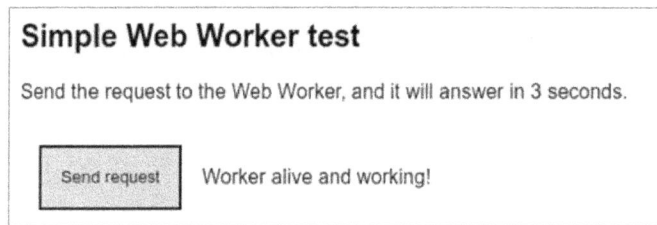

Figure 8.6 – The test application dispatches a command to the worker and shows the result

To make this happen, in our App.vue component, we import our web worker service and send our request with the command string as the name of the function in the service we want to execute. For this example, add the following code:

```
import webWorker from "./services/WebWorker.js"
webWorker.request("test").then(data=>{...}, err=>{...})
```

These simple lines of code to create and manage a web worker provide your application with a considerable increase in computational power and performance. Now that our bases are set, it is time to do something more significant with our service worker. Let's make it access our local database and the network.

Accessing IndexedDB with DexieJS in the web worker

IndexedDB is a very powerful key-value database; however, the native implementation provides an API that is rather hard to handle. The actual recommendation is not to use it but, instead, work with it through a framework or library. The database engine is fast and very malleable, so multiple libraries have built upon its foundation and recreated functions and features not present originally. Some libraries even mimic SQL and document-based databases. Some available and free-to-use libraries are the following:

- **DexieJS** (`https://dexie.org/`): A very fast and well-documented library that implements a NoSQL document-based database.

- **PouchDB** (`https://pouchdb.com/`): A database that mimics the functionality of Apache's CouchDB and provides built-in synchronization with remote servers.

- **RxDB** (`https://rxdb.info/`): This is a database that implements the reactive model. It also supports replication to CouchDB.

- **IDB** (`https://www.npmjs.com/package/idb`): This is a light wrapper implementation on top of the IndexedDB API, with some changes to improve its usability.

Depending on your requirements for local storage, these or other options will suit you well. We will use DexieJS for this example as it is well documented and boasts impressive speeds for bulk operations. We will expand our previous example and create a single-component mini-application to store, retrieve, delete, and view notes. This covers very basic **create, read, update, and delete** (**CRUD**) operations. When you run the example code, it will look something like this:

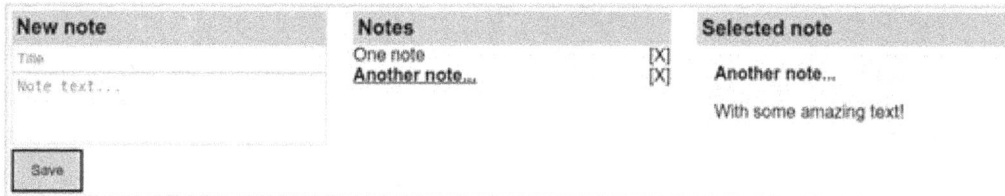

Figure 8.7 – A single-component CRUD example

In this example, you can create new notes, view what was saved before (this will be persistent based on the domain), select them to view the text, and also delete them. All the operations will be resolved in the web worker. Let's include Dexie in our application using npm:

```
$ npm install dexie
```

Next, let's create our example component application:

/src/components/DbNotes.vue

```
<script setup>
import webWorker from "../services/WebWorker"                      //1
import { ref } from "vue"
const _notes=ref([]),_note=ref({}),_selected=ref({})              //2
```

```
loadNotes()
function saveNote(){                                          //3
    if(_note.value.title && _note.value.text){
        webWorker
            .request("addNote", JSON.stringify(_note.value))
            .then(id=>{loadNotes()},err=>{...})
            .finally(()=>{_note.value={}})
    }
}
function deleteNote(id){                                      //4
    WebWorker
        .request("deleteNote", {id})
        .finally(()=>{loadNotes()})
}
function openNote(note){_selected.value=note;}               //5
function loadNotes(){                                         //6
    webWorker
        .request("getNotes",[])
        .then(data=>{_notes.value=data;},
              ()=>{_notes.value=]})
}
</script>
<template>
<div>
    <section>
        <h3>New note</h3>
        <input type="text"
               v-model="_note.title"
               placeholder="Title">
        <textarea v-model="_note.text"
                  placeholder="Note text..."></textarea>
        <button @click="saveNote()">Save</button>
    </section>
    <section>
        <h3>Notes</h3>
        <div v-for="n in _notes" :key="n.id">
            <a @click="openNote(n)">{{ n.title }}</a>
            <a @click="deleteNote(n.id)">[X]</a>
        </div>
    </section>
    <section>
        <h3>Selected note</h3>
        <strong>{{ _selected.title }}</strong>
        <p>{{ _selected.text }}</p>
    </section>
```

```
</div>
</template>
```

The preceding file has been stripped of styles and other layout elements, so we can focus on the active parts of the code that implement the operations we are learning about. We start by importing our service class to handle the web worker in line *//1* and create a few internal reactive variables in line *//2*. We will use _notes to hold the full list of notes as extracted from the database, _note as a placeholder to create new notes, and _selected to display a note clicked on from the list. You can find the CRUD operations in each function (lines *//3* to *//6*) and will notice that they are very similar other than handling UI reactive elements. They just gather the necessary information to create a request to the web worker and then apply the result. However, notice how in the saveNote() function, when it is time to pass the object that describes our new note, we are stringifying the Vue reactive value. This is because the proxy implementation that Vue uses to handle reactivity is not serializable, so unless we create a copy of the plain object or apply other similar techniques to extract the values, the web worker communication will fail and throw an error. An easy way to make sure the data object is provided as a clonable object is to convert it into a string with JSON.stringify(_ note.value) as in our code (you could also create directly a clone, with JSON.parse(JSON. stringify(_note.value))). You need to keep in mind how the information will be sent so that it can be properly processed at the receiving end of the web worker. This will become apparent now when we see dbService.js in the worker:

./src/webworker/services/dbService.js

```
import Dexie from "dexie"
const db=new Dexie("Notes")                                    //1
db.version(1).stores({notes: "++id,title"});                   //2
const service={
addNote(note={}){                                              //3
  return new Promise(async (resolve, reject)=>{
    try{
      let result_id=await db.notes.add(JSON.parse(note))       //4
      resolve({id:result_id})
    }catch(err){reject({})}
})},
getNotes(){
  return new Promise(async (resolve, reject)=>{
    try{
      let result=await db.notes.toArray();                     //5
      resolve(result)
    }catch{reject([])}
})},
deleteNote({id}){
  return new Promise(async (resolve, reject)=>{
    try{
```

```
        await db.notes.delete(id)                                    //6
        resolve({})
    }catch{reject({})}
})}}
export default service;
```

To use Dexie, we first import the constructor as in line //1, and we create a new database with the name Notes. Before we can actually use it, we need to define the version and a simple schema of tables/collections with the fields that will be indexed. This is what happens in line //2, where we define the notes collection with two indexed fields: id and title. These indexed fields are passed as a string, comma-separated by field names. We also included a double plus sign as a prefix for the id field. This makes the field auto-generated by the database and auto-incremented with each new record.

> The next significant function, addNote(), adds the record to the notes collection. Since we are passing data serializing an object as a string in our component, in line //4, we need to parse the string to recompose the object.

In the getNotes() function, we just retrieve all the elements from the collection and use the toArray() method provided by Dexie, which will convert it into a JavaScript array (line //5). This way, we can return it directly as our result to resolve the promise.

A final note on this code is on the deleteNote() method: on line //6, we are not capturing the result of the asynchronous operation. This is because this operation does not return a usable value. In this case, this operation will always resolve unless a database engine error interrupts the execution.

It is important to keep in mind that errors on the web worker will not affect the parent process, and any operations in that process will be unaffected.

Now that we have our service in place, it is time to slightly modify the web worker index file. Add the following lines:

./src/webworker/index.js

```
import dbService from "./services/dbService";
const services=[dbService, testService];
```

No other change is necessary for this file. As we can see, it does not take much to implement CRUD operations on the web worker. Even though these can be done in the parent process, and there is a small penalty from the interprocess communication, the benefits in performance are considerable and well worth the effort. Especially if our application includes what should be background processes, such as synchronization with a remote server, these should be done by a web worker. Let's see next how we can access the network and consume a **Representational State Transfer API (RESTful API)** from a worker as well.

Consuming a RESTful API with the web worker

One of the most common applications of network APIs today in web development is through the implementation of a RESTful API. It is a protocol where each communication is stateless and representative of the type of action required at the destination. The HTTP protocol used on the web provides a perfect match for this type of API, as each network call exposes a method that identifies the type of operation required:

- The GET operations retrieve data and files
- The PUT operations update data
- The POST operations create new data on the server
- The DELETE operations erase data on the server

It is easy to see how these methods match CRUD operations, so by making the appropriate network call, the server knows how to process the data received at the proper endpoint. There are many standards used to format the data sent between endpoints. In particular, one of the most common ones is the JSON format, which we so conveniently use in JavaScript.

Handling asynchronous calls with the native implementation in the browser is, at the very least, cumbersome but not impossible. The recommendation, for practicality and security, is to use a well-known library such as **Axios**. To install the library, we need to run the following command from the terminal:

```
$ npm install axios
```

After a few moments, the library will be installed into our project as a dependency. The library provides very handy methods to launch network calls for each HTTP method. For example, `axios.get` makes a GET request, `axios.post` makes a POST request, and so forth.

We will implement a simple service for our learning exercise to make network calls to a remote server from within our web worker. For simplicity, we will only create two methods:

./webworker/services/network.js

```
import axios from "axios"
axios.defaults.baseURL="http://localhost:3000"

const service={
   GET(payload={}){
     return new Promise((resolve, reject)=>{
       axios
         .get(payload.url,{params:{data:payload.data}})
         .then(result=>{
             if(result.status>=200 && result.status<300){
```

```
                resolve(result.data)
            }else{reject()}
        }).catch(()=>{reject()})
    })},
  POST(payload={}){
    return new Promise((resolve, reject)=>{
      axios
        .post(payload.url,{data:payload.data})
        .then(result=>{
          if(result.status>=200 && result.status<300){
              resolve(result.data)
          }else{reject()}})
        .catch(()=>{reject()})
    })}}
  export default service;
```

This service is rather simple. In a production application, it would be middleware to serve other services. This example implements only two methods to match the corresponding HTTP request methods. Notice that they are extremely similar, only changing the name of the method and the signature for some of the parameters. The first parameter is always the endpoint (URI) to connect. The second parameter is either data or an object with options. I refer you to the official documentation for how to handle each specific request and handle edge cases (`https://axios-http.com/docs/intro`).

It is worth noting that at the beginning of the file, we set the default domain for all other network calls. This way, we avoid repeating it in each call. We can easily set specific HTTP headers and options with this library, such as **JSON Web Tokens**, for authentication, as we covered in *Chapter 5, Single-Page Applications*, when we mentioned different authentication methods.

To include this service in our web worker, we import it and add it to our `services` array as we did before. Modify the beginning of this file so it looks like this:

./webworker/index.js

```
import netService from "./services/network"
const services=[dbService, netService, testService]
```

With this new inclusion, our web worker is now ready. We now implement a single component to test the communication, and it will look like this:

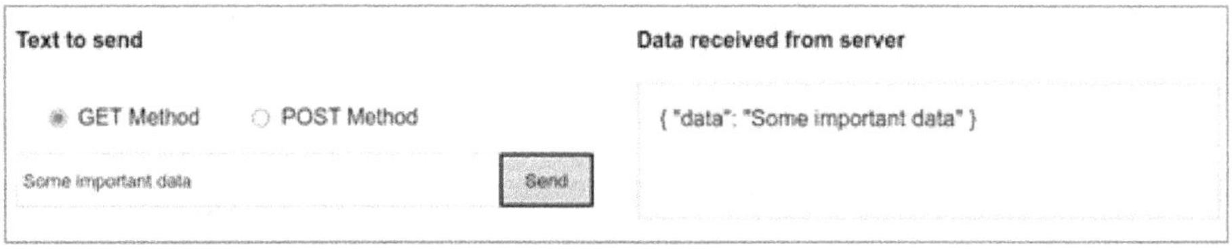

Figure 8.8 – A simple test, where the server mirrors back the information sent

Our component will let us select the method for the HTTP request (GET or POST) and will send some arbitrary data. The test server will just mirror the data received back to the client, where the component will present it on the screen. The implementation is quite straightforward:

./src/components/NetworkCommunication.vue

```
<script setup>
import webWorker from "../services/WebWorker"
import { ref } from "vue"
const
    _data_to_send = ref(""),
    _data_received = ref(""),
    _method = ref("GET")

function sendData(){
    webWorker
        .request(_method.value,                              //1
            {url:"/api/test", data: _data_to_send.value})
        .then(reply=>{_data_received.value=reply},
        ()=>{_data_received.value="Error"}
        })
}
</script>
<template>
    <div>
        <section>
            <h4>Text to send</h4>
            <div>
                <label>
                    <input
                        type="radio"
                        value="GET"
                        name="method"
                        v-model="_method">
                    <span>GET Method</span>
                </label>
                <label>
                    <input
                        type="radio"
                        value="POST"
                        name="method"
                        v-model="_method">
                    <span>POST Method</span>
                </label>
            </div>
```

```
                <input type="text" v-model="_data_to_send">
                <button @click="sendData()">Send</button>
        </section>
        <section>
            <h4>Data received from server</h4>
                {{ _data_received }}
        </section>
    </div>
</template>
```

In this component, we import the `webWorker` service and declare three reactive variables to send and receive data and one to hold the selected method to make the request. Our simple test server will receive the request and will just mirror back the data that we submit. We will see later how to create this simple server using Node.js.

In the template, the user can choose the type of request to send (`GET` or `POST`), a choice that we save in the `_method` variable. We use this value as the command passed to the worker in line //1. We pass the data as a member object as a payload. When this promise resolves, we save the value from the reply in the `_data_received` variable. The rest of the source code should be trivial to understand at this point, as it deals mainly with the template and presentation of the information on the screen. Before we end this chapter, let's take a look a how the test server can be implemented using Node.js.

A simple NodeJS server for testing

To test our network communications, it seems fitting that we implement a small server using Node.js to implement the endpoints that we are testing. In a separate directory from our Vue application, open a terminal window and enter the following command:

```
$ npm init
```

The command-line wizard will ask you a few questions to create the `package.json` file that represents a Node.js application. When it is done, run this command to install the **Express.js** dependency, which will give us a framework to create a web server:

```
$ npm install express cors
```

Once the process completes, create an `index.js` file with the following code:

./server/index.js

```
const express = require("express")              //1
const cors=require("cors")                      //2
const app=express()                             //3
const PORT=3000
app.use(cors())                                 //4
```

```
app.use(express.json())                                //5
app.get("/api/test", (req, res)=>{                     //6
    const data=req.query                               //7
    res.jsonp(data)                                    //8
})
app.post("/api/test", (req, res)=>{
    const data=req.body                                //9
    res.jsonp(data)
})
app.listen(PORT, ()=>{                                 //10
    console.log("Server listening on port " + PORT)
})
```

With these few lines of code, we can start a small server that receives and responds with JSON data. We import the express constructor (line //1), and a plugin (line //2). This is important so that we can access this server from any domain (origin). The **Cross-Origin Request Sharing (CORS)**serves to circumvent the security measure that servers implement to prevent serving requests from other sources (origins) than their own. It has to be enabled to allow requests from other origins. After we create our server application (line //3), we pass the plugin (line //4). We also pass another plugin (line //5), to make the server identify and reply to network calls with JSON objects. We then proceed to create two endpoints, one for the GET requests (line //6) and another one for a POST request. The first parameter is the URL where the server will be listening for calls. In this case, they are the same, as the only difference will be the type of request method. This is standard practice.

Each endpoint receives as the last argument a callback function with at least two parameters: req (request) and res (response). These objects contain methods and information about the request received and the necessary methods to create a response to the client.

For the GET requests, the data received is passed attached to the URL as a "query string", so in order to access it, *Express* wraps it nicely as an object in the request.query field (line //7). Since we are just replying with the same data received, in line //8, we use the res(ponse) object to create a padded JSON reply with the same data object. We do this because we consider that we could receive calls from any domain (because we enabled CORS) and want to ensure the reply is fully understood. **JSON with Padding (JSONP)** is a method to send the response using a different method. We don't need to worry about this, as both ends (sender and receiver) are handled by the Express server and the Axios client.

In the post method, the difference is that the data is contained in the body of the message (line //9), hence the different treatment. Finally, the server starts listening on the designated port (line //10). We can now access the server at http://localhost:3000, which is the address that we configured in our network service as the default for Axios.

With the implementation of this server, we can now have a full test of all the parts of the system.

Summary

In this chapter, we reviewed some very important concepts to fundamentally improve the architecture and performance of our application. Web workers is an amazing technology that allows web applications to take advantage of modern hardware architectures and modern operating systems. From a fixed point of view, multithreading using web workers involves little additional effort and complications, and the gains are highly rewarding. We also saw how to make use of workers to access network services as well as the local persistent database provided by the browser (IndexedDB). We learned about two more design patterns to implement a scalable architecture for our application and tested the concepts and implementations through simple components and services. The use of web workers marks a significant difference in the performance and execution of a well-designed web application. In the next chapter, we will look into tools and techniques to test our code automatically, ensuring the individual parts comply with their intended purpose to match our software specifications and requirements.

Review questions

- What limitations does JavaScript have that can compromise the performance of a web application?

- What are web workers? What are their limitations?

- How can Vue applications communicate with web workers?

- What are the benefits of using a design pattern such as the Business Delegate to work with web workers?

- What can you change in the example code to manage multiple web workers instead of just one? When would this be advisable, in your opinion?

9

Testing and Source Control

The success of our application depends on many factors, beyond the quality of our code organization or patterns. Moreover, the very nature of software implies that there will be changes during and after development, changes in the requirements, the scope, and so on. With each feature developed, an item of complexity is introduced into the software, creating relationships and dependencies. New inclusions may disrupt these connections and introduce breaking changes, bugs, or even completely disable the system. The solution for this problem is to keep track of code changes and conduct tests on the application to identify problems and ensure as much as possible that the system complies with the desired software attributes and satisfies the requirements.

In this chapter, we will cover the following:

- Different approaches to testing and the concept of **test-driven development** (TDD)
- Installing a test suite (Vitest) and test tools (Vue Test Utils) for our project
- Creating and running tests on an existing project for synchronous and asynchronous code
- Testing our components by simulating user interactions
- Installing and managing our source code using Git and online repositories such as GitHub or GitLab

The concepts in this chapter are introductions to important professional skills for a developer to ensure the delivery of good-quality software. Often, these tasks are left aside or relegated as an afterthought. However, the lack of them may lead to expensive mistakes and lengthy overwork as the software complexity grows. For non-trivial applications with more than one developer involved, nowadays, it is hardly possible to conceive a project that does not use some of these tools.

In this chapter, we will focus on **unit testing** and the tools provided by the Vue team to perform it.

Technical requirements

This chapter does not have additional requirements to those of previous implementations of code examples. The final source code can be found in the official repository for this book at `https://github.com/PacktPublishing/Vue.js-3-Design-Patterns-and-Best-Practices/tree/main/Chapter09`.

Check out the following video to see the Code in Action: `https://packt.link/UqRIi`

What are testing and TDD

Testing is the process to verify that the software is doing what it is supposed to do, according to the requirements of the project. It involves the manual or automated execution of tools to evaluate and measure different properties and attributes of the software, identify errors and bugs, and provide feedback for developers to take action to correct them. There are many different approaches and types of tests to be performed, such as the following:

- **Unit testing**: This is where relevant units of the source code are validated against a series of inputs and outputs. It is often automated.

- **Integration testing**: All the components of a system are verified together as a group, looking for errors and bugs in the resulting integration, communication, and so on.

- **End-to-end testing**: This involves a complete validation of the application simulating real-world use, interacting with databases, network scenarios, and so on. It can be performed with automated tools that simulate human interaction, and manual testing using real-life users.

These types of testing are just a small sample of this discipline, as there are hundreds of possible tests to apply to the software. Large companies may have entire testing teams dedicated to ensuring the quality of the software. Usually, the more complex the software, the more complex the testing may be. In practice, the testing plan can be as complex as the development plan itself. As mentioned in the introduction, we will focus on the official tools provided by the Vue team for this task.

Testing can be done before, during, after, or in parallel with the development. **TDD** is a discipline that places the burden of testing as early in the project as possible, even before the actual coding begins, with the objective to match the requirements. It involves the following steps:

1. Write a test case, based on the requirements and design of the application, with key inputs and expected outputs.
2. Run the test, which should fail (as there is no code written yet).
3. Write the actual code to be tested (a function, Vue component, etc.).
4. Run the test against the created code. If it fails, refactor the code or design.
5. Start again with a new test case for the next unit.

This process is repeated, and it is expected to provide developers with a significant reduction of "bugs" and errors and help them focus on the requirements. This does incur an overhead of effort early in the project, as opposed to refactoring, when the tests are performed toward the end.

TDD has become popular in some teams and with some frameworks, and it is supposed to help developers improve their own code as they now acquire a "testing" mindset. However, there are no specific studies made to confirm this, but practitioners of this discipline do report that it has improved their code and design. This, of course, begs the question: what needs to be tested, and how can we streamline the task into our workflow? That is the topic we will discuss next.

What to test

A key factor in the success of a good testing plan and implementation is deciding what to test. It is not possible to test the full universe of possibilities or 100% of the components and interactions in a project when considering internal and external factors. Even the attempt to fully cover all possibilities would be incredibly expensive and practically impossible. Instead, the focus needs to be on the real possibilities of what can be tested within our time and budget constraints, by carefully selecting the non-trivial elements that "make or break" our project requirements. This is often not an easy task.

When it comes to Vue applications, we need to focus on crucial services and components that perform key operations. We need to test the following:

- **Services**: Self-contained functions, both synchronous and asynchronous. Functions that don't return a value but perform logical procedures will serve a different kind of testing than what we will see here. These will involve mocking network communications or database calls, application policies, and so on. However, the principles for testing these are similar.

- **Components**: We need to test inputs (props) and outputs (events and HTML). Higher-level components that group other components to perform a workflow or business logic can also be tested in the same way (props, events, and HTML rendered). However, these also will need other types of testing, such as end-to-end testing.

We can write our own functions and tools to perform tests, but apart from some edge cases, the obvious recommendation is to use stable test suites and tools. In our case, for Vue, there are official resources provided by the same team, called **Vitest** and **Vue Test Utils**. Using a testing suite/library has many benefits, akin to the use of a framework or library in the "regular" development of an application. Perhaps one of the major benefits has to do with **DX**, or **Developer eXperience**, as they streamline and make the development process easier or lighter in the best case. Let's learn how to apply these tools in our workflow by going through an example application, which we will address in the next section.

Our base example application

It is best to understand the discipline of testing and learn about the tools by applying them to a real project through practice. As a learning exercise, we will first take a running application based on one of the examples presented in *Chapter 2, Software Design Principles and Patterns*. We will build a **Fibonacci calculator** and install the Vitest test suite and Vue Testing Utils to the project. Later, we will explain what would change in this approach when applying the TDD discipline.

The code for this application can be found in the repository for this chapter. Once downloaded, you need to execute the following command to install the dependencies:

```
$ npm install
```

Then, to run the application, you must run the following:

```
$ npm run start
```

When the server is ready, loading the site in your web browser should present you with an application like this:

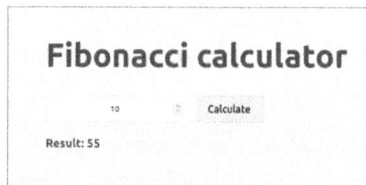

Figure 9.1 – The example application with a Fibonacci calculator

The design of this application has been made with the purpose of learning the basics of testing functions and components, so it is very basic but sufficient. We are presented with one service file (`/src/services/Fibonacci.js`) and three components: `App.vue`, `FibonacciInput.vue`, and `FibonacciOutput.vue`.

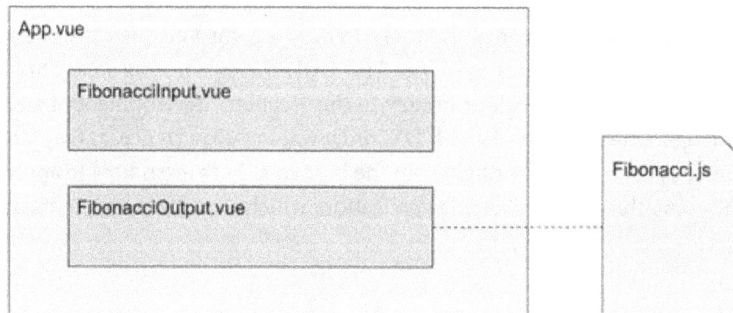

Figure 9.2 – Components and service for the application

Our application-level component, App.vue, receives from FibonacciInput.vue a positive integer number through an event, which passes as a prop input to FibonacciOutput.vue. This component uses the Fibonacci.js service to calculate the respective Fibonacci number corresponding in the series and present it to the user. As simple as this application sounds, it gives us basic examples to create tests for the most common cases, which will give us a solid start. It is now time to install our test suite.

Installation and use of Vitest

Vitest (https://vitest.dev/) is a test suite, meaning that it provides, out of the box, a set of tools and a framework to perform tests in our code. Being developed by the Vue and Vite teams, it integrates seamlessly with Vite, even sharing the same configuration and respecting each other's organization. *Vitest* can be selected during the original creation of a Vue project by selecting the proper option during the creation wizard – a task that will add a /src/__tests__ folder, some examples, and a few extra entries in our package.json file. But all this boilerplate can be a bit confusing unless we have previous experience in this area. Instead, we start from an already created project, so we will install Vitest as a development dependency – a task that will give us an insight into how it works and is organized.

Install Vitest from the command line at the root directory of the project with this command:

```
$ npm install -D vitest
```

The package manager will take some time to include Vitest and all necessary dependencies but won't modify our source code or organization. For convenience, we will use npm as well to run our tests, so we need to open our package.json file, and in the scripts section, enter the following lines so the section looks like this:

```
"scripts": {
    "start":"vite",
    "build": "vite build",
    "preview": "vite preview",
    "test": "vitest",
    "test:once": "vitest run",
    "test:coverage": "vitest run --coverage"
}
```

At this point, we can now test our test suite:

```
$ npm run test
```

After you run that command, you will be graciously greeted with a red message explaining that the tests have failed. Perfect. This is what it should do, as we do not have any tests yet! So, let's add them. We will start by testing our Fibonacci.js service.

Vitest allows us to write our test functions in independent files or in-source, meaning that we place them alongside our component's JavaScript. There are benefits and trade-offs with both approaches, but to start, we will place our test code in independent files, one for each service and component. In this way, we place these files in their own directory, which by convention can be either `/src/tests` or `/src/components/__tests__`, but they could also be placed alongside the Single File Components or with the services. Vitest will scan the entire source folder for the test files. Even though we can be very creative when placing these files, we will place them in `/src/test` to keep things neat and tidy. There is another convention to adhere to, which is that each test file must have the same name as the file being tested, plus the `.spec.js` or `.test.js` extension. Vitest uses this convention to identify and run the tests in an organized manner. So, in our case, our `Fibonacci.js` service will have its testing counterpart in `/src/tests/Fibonacci.test.js`. Go ahead and create that file, and enter the following lines:

/src/tests/Fibonacci.test.js

```
import { describe, expect, test } from "vitest"
import { Fibonacci, FibonacciPromise } from "../services/Fibonacci.js"
```

In the first line, we import three functions from Vitest, which are the foundation of all our testing, and the ones that we will be using most often. Here is what each one does:

- `describe(String, Function)`: This function groups together a number of tests, and Vitest will report the test group by using the description given as the first parameter. The second parameter is a function, where we will run the tests with the `test()` function.

- `test(String, Function)`: The first parameter is a description of the tests encompassed in the second parameter, which is a function. The test will "pass" if no errors are thrown within it. This means that we can write our own test logic and tools following this condition and throw a JavaScript error when the validation fails. However, there is a simpler approach...

- `expect(value)`: This is the function that performs the "magic" of testing. It receives, as a unique argument, a single value or a function that resolves to a single value. The result of `expect()` is a chainable object that exposes many different and almost language-natural assertions (comparisons, validations, etc.) to perform on the argument value. Under the hood, it uses the Chia syntax to a certain extent and is also compatible with other test suites, such as Jest – for example, `expect(2).toBe(2)`. A full list of all the possible assertion methods can be found in the official documentation here: `https://vitest.dev/api/expect.html`.

In the second line of the test file, we directly import the two functions contained in the service: `Fibonacci()` and `FibonacciPromise()`. We need to import each function that we want to test, and then create as many test groups as necessary for each one. Let's start with the self-contained `Fibonacci()` function by adding the following test group:

```
describe("Test the results from Fibonacci()", () => {
  test("Results according to the series definition", ()=>{
    // Expected values as defined by the series
    expect(Fibonacci(0)).toBe(0)
    expect(Fibonacci(1)).toBe(1)
    expect(Fibonacci(2)).toBe(1)
    expect(Fibonacci(3)).toBe(2)
    // A known value defined by calculation of the series
    expect(Fibonacci(10)).toBe(55)
  })
})
```

We start by creating a test group with `describe()` and create inside the passed function as many tests as needed. Inside each `test()` function, we can create as many assertions as needed, but it has to have at least one. Notice how we are executing the function from the service with different arguments, and then asserting them to the expected value as defined in the numerical series. In this case, we are using `.toBe()` to test equality, but in the same way, we could be testing strings, objects, types, and so on, using other assertions, such as `.not`, `.toEqual`, `.toBeGreaterThan`, etc. There are more than 50 assertion methods defined in the documentation (`https://vitest.dev/api/expect.html`). Take some time to review them, and remember that these are chainable, so you can make more than one assertion at once.

After saving this file, you can run the test again:

```
$ npm run test
```

You should receive a few messages in green, indicating the number of tests performed and whether they passed or not. In the case that one raises an error, it will be pointed out in red letters using the descriptive text and line where it occurred. That is a sign to start refactoring the code (assuming the test function and assertion were properly and correctly written; otherwise, you get a false positive!).

In the case that no assertion method works for a particular edge case, you can create inside `test()` your own logic in plain JavaScript and throw an error when the validation fails. For example, these two code snippets are equivalent:

```
// Using expect
expect(Fibonacci(10)).toBe(55);

// Using your own logic
let result=Fibonacci(10);
if(result!=55) throw Error("Calculation failed");
```

Even though this example is trivial, it is easy to see how the first case, using `expect()`, results in a better developer experience, as it is succinct, elegant, and easy to read.

> **Vitest is still running!**
>
> Perhaps you have noticed that running `npm run test` does not end the execution of the script once the tests have terminated. Just like with a developer server, Vitest keeps waiting for changes to occur to the source code or test files and automatically reruns all the tests for you. If you want to run the tests only once, use `npm run test:once` or `vitest --run` to flag Vitest to run the tests only once and then exit.

Special assertion case – fail on purpose

All the previous assertions thus far have been made using the "positive" approach that a function will return what is expected. Using the "negative" approach in testing is to make sure that a function will fail when it is supposed to. For example, the Fibonacci series is not defined for negative numbers, so any calculation should not return a value but should throw an error. In these cases, we need to wrap the execution of the function in another function, thus encapsulating it to test the assertion against a thrown error. This would be the equivalent of using a `try..catch` block in plain JavaScript to avoid terminating the execution of the script when an error occurs. For example, executing `Fibonacci(-5)` should throw an error, so we will write our test case as this:

```
test("Out of range, must fail and throw an error", ()=>{
    expect(()=>Fibonacci(-5)).toThrow()
})
```

The preceding assertion will work as expected, without interrupting the testing process.

Special assertion case – asynchronous code

Another special case to keep in mind is asynchronous code, such as network calls, promises, and so on. In this case, the solution is to use `async..await`, not on the function but on `expect`. For example, to test the `FibonacciPromise()` asynchronous function, we would write a test like this:

```
test("Resolve promise", async ()=>{
    await expect(FibonacciPromise(10)).resolves.toBe(55)
})
```

Notice how we are applying the `async` syntax to the entire test function, and `await` to the `expect()` function. We also need to use the `.resolves` assertion to indicate the successful resolution for the value to validate. If we needed to test a `Promise` rejection, we would use `.rejects` instead of `.resolves`.

With this, we have covered the majority of tools and test approaches to get us started in unit-testing our plain JavaScript functions. However, all these tests are executed using **Node.js** (the server version of JavaScript), not on the browser where our Vue components will be executed. In Node.js, there is no **DOM** or **Windows** object, so we don't have any HTML… so how do we test our **Single File Components**?

The answer is to provide Vitest with a simulated DOM where we can mount our components and run tests as if it were a browser window. Here is where the Vue Test Utils tools come into play.

Installation of Vue Test Utils

As of now, Vitest provides us, out of the box, with tools to test plain JavaScript functions, classes, events, and so on. To test our Single File Components, we need additional resources, and these are provided to us again by the official Vue team in the form of **Vue Test Utils** (`https://test-utils.vuejs.org/`). To install them, run the following command:

```
$ npm install -D @vue/test-utils
```

Once the installation has completed, we need to update our `vite.config.js` file to include the environment where the components will be tested, meaning a browser context. Modify the configuration file so it looks like this:

```
export default defineConfig({
    plugins: [vue()],
    test:{environment:"jsdom"}
})
```

Vitest and Vue Test Utils both integrate seamlessly with Vite, to the point that they share the same configuration file. You can now run the test suite, and Vitest will attempt to download and install any missing dependencies on the first run after these modifications. If for some reason the installation of jsdom does not happen automatically, you can install it manually with this command:

```
$ npm install -D jsdom
```

Now, with these changes, we are ready to start our first component tests. Let's start creating a file to test our `FibonacciOutput.vue` component, as it is the simplest, we have in our application. Create the following file in the test directory with this code:

/src/tests/FibonacciOutput.test.js

```
import { describe, expect, test } from "vitest"
import { mount } from "@vue/test-utils"                          #1
import FibonacciOutput from "../components/FibonacciOutput.vue"   #2

describe("Check Component props and HTML", () => {
    test("Props input and HTML output", () => {
        const wrapper = mount(FibonacciOutput,
            { props: { number: 10 } })                           #3
        expect(wrapper.text()).toContain(55)                     #4
    })
})
```

The preceding code is not that different from a basic unit test as we have done before, but it does some things a bit differently. In line #1, we import a function from the Vue Test Utils library that allows us to "mount" our component in a test environment simulating a browser window with Vue 3. In line #2, we import our component in the usual way, and then proceed to write our test group as before. The difference here is in line #3. We use the `mount` function to create our live component by passing it as the first argument and, as the second, we pass an object with properties that will be applied to the component. In this case, we are passing the `number` prop with a value of `10`. The `mount` function will return a wrapper object representing our component, exposing an API that we access to perform our assertions. In this case, in line #4, we are checking that the plain text rendered by the component contains the value 55, which we will find to be true when the test is run. It is by using this wrapper object that we can access the component properties, events, slots, and rendered HTML by accessing the proper methods. We will only discuss a few in this chapter, but a full list is available in the official documentation at `https://test-utils.vuejs.org/api/#wrapper-methods`.

This short example gives us a template to write our tests, but now we move to a more complex example to test our `input` component. In the test directory, create the following file:

/src/tests/FibonacciInput.test.js

```
import { describe, expect, test } from "vitest"
import { mount } from "@vue/test-utils"
import FibonacciInput from "../components/FibonacciInput.vue"

describe("Check Component action and event", ()=>{
    test("Enter value and emit event on button click", ()=>{
        let wrapper=mount(FibonacciInput)                         #1
        wrapper.find("input").setValue(10)                        #2
        wrapper.find("button").trigger("click")                   #3


        // Capture the event parameters
        let inputEvents=wrapper.emitted("input")                  #4
        // Assert the event was emitted, and with the correct value
        // Each event provides an array with the arguments passed
        expect(inputEvents[0]).toEqual([10])                      #5
        // or
        expect(inputEvents[0][0]).toBe(10)                        #6
    })
})
```

This final example starts in the same way as before, by importing the functions that we will use to describe the tests, mount our component, and the component itself. Our purpose here is to simulate to a certain degree the user interaction with the component by entering a value in the `input` field, clicking the button, and then capturing the event and the value passed programmatically. We will rely on the methods just like before. We start in line #1 by mounting our component and creating the wrapper. Notice that, this time, we are not passing any options, as we don't need them. In line #2, we use the wrapper's `find()` method to locate an `input` element and set a value of `10`. The `find()` method retrieves elements using a string with the same syntax as `querySelector` in a browser window. The returned object is a wrapper around the element, which again exposes methods for users to interact with it – in this case, `.setValue()`. Using a similar logic, in line #3, we also locate the button and trigger the `click` event, which will emit the `input` event in our component. Notice how easy it is in lines #2 and #3 to manipulate our component. In this way, we can access and interact programmatically with it, much like it could happen in an end-to-end test. We could, in theory, create our end-to-end tests using this tool, but there are better options, such as **Cypress** (`https://www.cypress.io/`), which work excellently with Vitest, giving us a great DX.

In line #3, we have clicked a button, which we know should emit an event. In line #4, we capture all the emitted events with the name `input`. The result is an array of wrapped events that we can use in our assertions, by referencing each event by its ordinal index. In this case, we only triggered one event, so in line #5, we pass that to our expected function as `inputEvents[0]`. However, notice that the assertion matches the output to an array, `[10]`, instead of the value we entered in line #2. Why is that? The answer is that each *event* has an undetermined number of arguments it could pass, so these are captured in an array. An equivalent notation is shown here in line #6, where we pass to `expect()` directly the value of the first element in the array of arguments, from the first event captured: `inputEvents[0][0]`. Then, we can directly validate the result to a value with `.toBe(10)`. Now, this approach may seem a bit convoluted and clumsy, having to refer to events and their values in such a way, but it is very powerful. Consider that we can, in one single line, assert a full array with a set of related values!

In these two files, we have now tested the input and output of our components and even validated the interactivity as expected. We have also learned how to retrieve elements rendered and access their properties. Any error thrown in these functions will invalidate the test and point us in the right direction, line, and comment on where to fix it. Placing tests in individual files is a very convenient alternative. However, Vitest also accepts in-source testing, which we will see next.

In-source testing

With in-source testing, we can indicate to Vitest to look into our JavaScript and Single Component Files for the tests to run, as opposed to specific files. These alternatives are not exclusive to each other, so we could have both active at the same time. The reason behind this is that, in some cases, a test case would benefit to be "close" to the original code that it is trying to assert. Such code must be placed at the end of our file following this format:

```
if (import.meta.vitest) {
    const { describe, test, expect } = import.meta.vitest
    //... Test functions here
}
```

Then, for Vitest to find this code in our files, we also need to modify the `vite.config.js` file to include the following:

```
export default defineConfig({
    test: {
        includeSource: ['src/**/*.{js,ts}'],
        // other configurations here...
    },
})
```

And finally, to eliminate the test code from the production build, we need to add the following before the bundling:

```
export default defineConfig({
    define: { 'import.meta.vitest': 'undefined' },
    // Other configurations...
})
```

With these changes, we can then include the tests at the end of our JavaScript files, with the benefits and trade-offs that this brings. For example, if there is an in-house service that is shared or used among projects, it could be a good idea to place the tests in the same file as opposed to duplicating them for each project.

Now that we have our tests in place, let's see two more benefits that we get from using Vitest: *coverage* and a live *web UI*.

Coverage

The concept of coverage is very simple, and it answers the question of how much of our code is covered by automated tests. We know that 100% coverage is only possible for small applications, as the same effort for large projects falls fast into the law of diminishing returns Vitest offers us a simple way to answer this question by running the `vitest -coverage` command. In our case, we have already set this option in our `package.json` scripts section, so we can run the following command:

```
$ npm run test:coverage
```

When the preceding command is run, if any dependency is missing, it will prompt us on whether we want to try to download and install it:

Figure 9.3 – Vitest prompts us to install missing dependencies for coverage

For our chapter code example, the coverage report should look something like this:

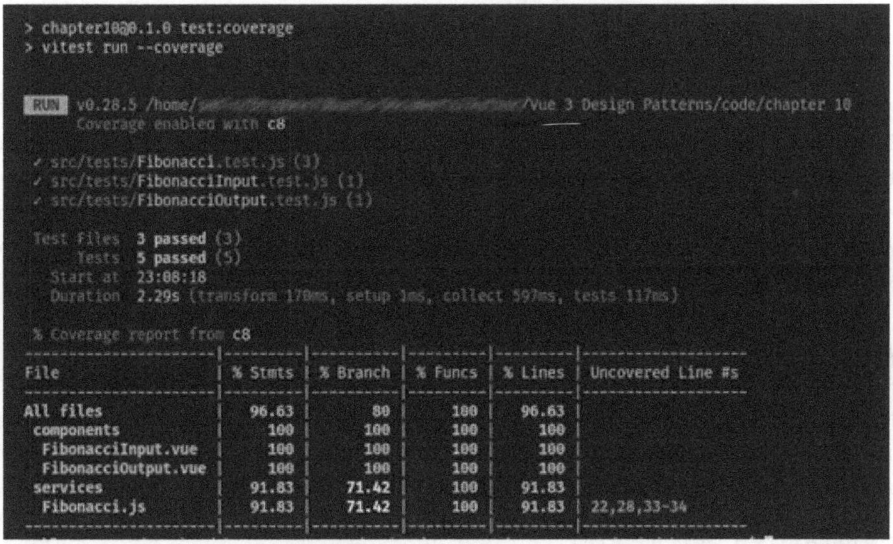

Figure 9.4 – Vitest coverage report example

It is possible to retrieve this information on a file (as json, text, or html) if we need to. For that, we just need to include a new line in our vite.config.js file:

```
test:{
    coverage: {reporter: ['text', 'json', 'html']},
    //...
}
```

The result of running the command again will be a website placed in a new directory called `coverage` at the root of our project. This static website provides for navigation and drills down in the report. In our example, it looks like this:

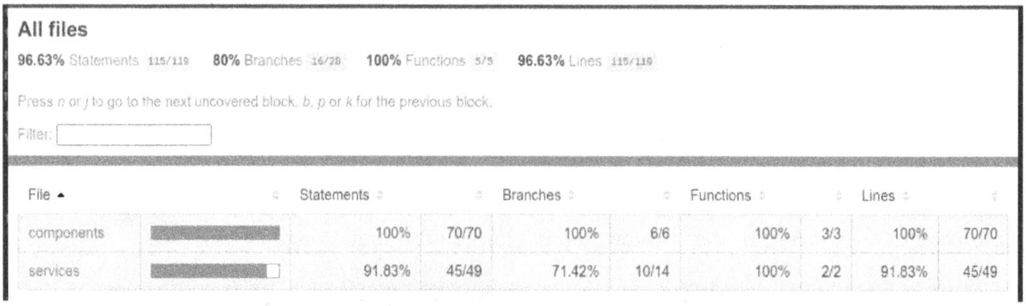

All files

96.63% Statements 115/119   80% Branches 16/20   100% Functions 5/5   96.63% Lines 115/119

Press *n* or *j* to go to the next uncovered block, *b*, *p* or *k* for the previous block.

Filter:

File ▲		Statements		Branches		Functions		Lines	
components		100%	70/70	100%	6/6	100%	3/3	100%	70/70
services		91.83%	45/49	71.42%	10/14	100%	2/2	91.83%	45/49

Figure 9.5 – Coverage HTML report

Depending on our needs, this simple tool may provide us with insight into our project that would be hard to find otherwise. The export to JSON file comes also very handy if we need to integrate our project with other reporting software or format. There is yet another alternative that may come in handy: Vitest also provides a web UI to view and interact with the tests in the form of a dashboard. We will see this next.

The Vitest UI

Since Vitest is based on Vite, it does make good use of some of its features, not only for live testing but also to provide a live development server displaying the tests in real time. To make use of this option, we only need to install the appropriate dependency as follows:

```
$ npm install -D @vitests/ui
```

Then, for convenience, we should add the following line in our `package.json` file so we can run the application using npm:

```
scripts:{
    "test:ui": "vitest --ui"
    // Other settings...
}
```

We can then run the server with this command line:

```
$ npm run test:ui
```

The development test server will start and provide us with an address to open in the browser. For our application, this looks like this:

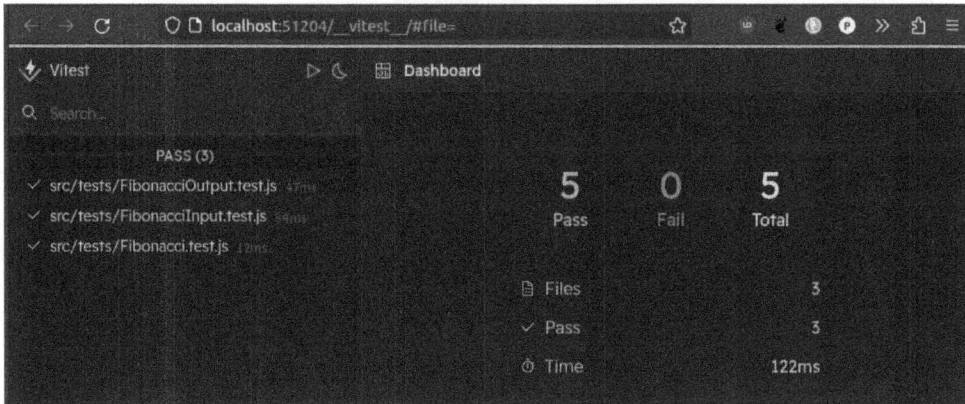

Figure 9.6 – Vitest UI dashboard

The web UI has also new possibilities to interact with the test cases, and even see the relationships between components and services in a graphical manner, all the way down to the test code.

Now that we have cleaned up our code and run tests, it is time to look into another tool to keep track of changes, a fundamental concept for today: source control with Git.

What is source control... and why?

Software development is a "human-intensive" discipline, meaning that it depends heavily on the creativity and involvement of the developer and their know-how. It is common to try different approaches to the same situation and write and rewrite code. Even the process of refactoring after testing implies making changes in the code. It is not an anomaly that during this process, we need to "go back" to a previous code when a change or approach didn't meet expectations. If we are constantly overwriting the same files... how do we keep track of what changed where? And by whom? Our own memory is not enough when time and complexity grow. Save files with different names? That would become impractical very soon. And what about combining source code from multiple developers? We can quickly see that managing the source code for non-trivial projects is a very important task in itself.

The historical solution to this early problem in computer science has been the creation of additional software in charge of keeping track of changes in the code, allowing a developer to go back on their tracks, and facilitating the chore of merging code from multiple developers to make a cohesive source code. The emerging discipline for this task is called **Source Control** (**SC**), and the software to implement it is called a **Source Control System** (**SCS**) or **Source Control Management System** (**SCMS**). There have been many, and still are many, different systems in use today, such as **Mercurial**, **Subversion**, **ClearCase**, **Git**, and **BitKeeper**. Each one has its trade-offs. In particular, Git is used today by most projects and developers around the globe. Statistics on the internet show different percentages for the most popular ones, but each one shows this trend. Because of this, it is important that we learn how to use Git, which is our next topic.

Source control with Git

Currently, the most popular SCS is Git, which was created by **Linus Torvalds,** who is also the creator of the Linux kernel. The story goes that the Linux kernel project used *BitKeeper* for source control, but the team hit many issues with the growing complexity and distributed nature of the development. Frustrated, Linus Torvalds decided to make his own SCS to solve the real-life problems they had... and it took him one weekend! (See `https://www.linux.com/news/10-years-git-interview-git-creator-linus-torvalds/.`) That was the humble beginning of Git, and from there, it became popular in the open source community as well as in the enterprise world.

Git is a distributed SCMS, simple and effective to use from the command line. It offers the following features:

- Creates and manages a **repository**, where it collects the source files and the history of changes for each one.

- Allows sharing projects by cloning **remote repositories** into local projects.

- Allows the project to be branched and merged. This means that you can have different copies of the same project with different code (a **branch**), switch between them, join them, and unify them by request (a **merge**).

- Synchronizes changes from a remote repository into a local copy (called a **pull**).

- Sends local changes to a remote repository (called a **push**).

Let's learn how to use Git by applying it to our current project for this chapter. Let's start by installing it in our system, so it's available for all our projects.

Installation on Windows systems

The easiest and recommended way to install Git on Windows systems is to download the installers from the official Git website at `https://git-scm.com/download/win`. Click on the version you want to use, according to your operating system (32- or 64-bit), and then run the installer following the directions.

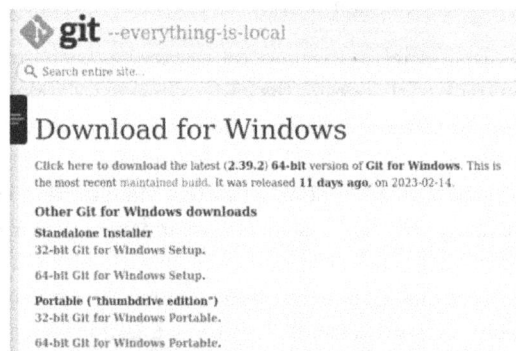

Figure 9.7 – Official Git installers for Windows

Once the installation is done, the command-line tools will be installed on your system so we can run them through a terminal. Also, if you are using a code editor such as Visual Studio Code, it will integrate the tools and provide you with a GUI to handle basic operations.

Installation on Linux systems

In Linux systems, the installation is done through the command line, using the distribution's package manager. The package name in (almost) all distributions is simply `git`. In Debian and Ubuntu systems, the installation can be run with the following:

```
$ sudo apt install git
```

However, in these distributions, there may not be the latest version, so if you need the latest stable release, you need to add the official PPA repository. In this case, run the following commands in order:

```
$ sudo add-apt-repository ppa:git-core/ppa
$ sudo apt update
$ sudo apt install git
```

The preceding commands will update your system dependencies and install (or upgrade) Git on your system. For a complete list of distributions and commands to install Git, please refer to the official documentation at `https://git-scm.com/download/linux`.

Installation on macOS systems

In macOS systems, there are different ways to install Git:

- If you have Homebrew installed, run `$ brew install git` in a Terminal
- If instead, you have MacPorts, run `$ sudo port install git` in a Terminal
- If you have installed Xcode, Git is included

For other alternatives, please check the official documentation at https://git-scm.com/download/mac.

Using Git

Regardless of which system you are working on, or the installation type you made, Git will be installed in your local path, so it can be executed from any terminal window. To verify the installation and version, run this command (does not require admin privileges):

```
$ git --version
```

At the time of writing, the current stable version is *2.39.2*. With this done, open a terminal window in the root folder of our project. To start using Git, we need to create a local repository with this command:

```
$ git init
```

After the execution is done, a new hidden directory will be created in the folder. You don't have to worry about it, as it will be managed by Git. If your File Explorer has deactivated the option to see hidden files, then you may not notice the creation. It is recommended that you have **Show/view hidden files** in your system activated.

Once we have created the repository, we can start using it. The steps to work with files usually includes the following stages:

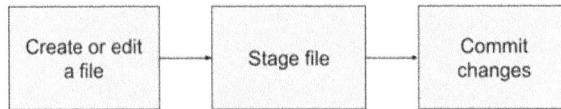

Figure 9.8 – Working stages of Git

Once we have files created or edited, the next step is to "stage" the files. This indicates to Git that it needs to keep track of changes and include the file in the next commit event. **Committing** is the act of moving those files/changes into the repository. If a file is not staged, it will not be included in the commit. To add a file, run the following command:

```
$ git add [filename1] [filename2]..
```

This will add the files, but it is quite verbose. Instead, if you want to add all changes in all files, run this instead:

```
$ git add .
```

This will come in handy in the first commit when the repository is initialized. After running this command, all the files will start to be tracked. However, we do not want to track everything in our root folder, so to exclude files or directories, we can use a special file named `.gitignore`. If you open this file in the example directory, you will find something like this:

/chapter 10/.gitignore

```
logs
*.log
npm-debug.log*
yarn-debug.log*
yarn-error.log*
pnpm-debug.log*
lerna-debug.log*
node_modules
dist
dist-ssr
```

```
*.local
.vscode/*
!.vscode/extensions.json
.idea
.DS_Store
*.suo
*.ntvs*
*.njsproj
*.sln
*.sw?
```

This is a plain text file that indicates to Git not to track the files and directories indicated in each line. You can also use wildcards such as an asterisk (*) and question mark (?) to include a match pattern. This is very useful, as there are parts of your code base that you don't need to track, such as the node dependencies and binary files (images and the like). Make sure to have this file in your directory before making a massive staging.

Once you have staged files, you can check them with this command:

```
$ git status
```

In the case of our example project, it will show something like this:

Figure 9.9 – First staging in Git

Notice how Git informs us also that we are in the `master` branch, and that there have not been any commits yet. The `master` branch is the main branch for our code and is created by default. This is a special branch that is used to keep the stable code of our applications. In tools such as GitLab and GitHub (we will talk about them later), these branches also trigger certain events once committed. For now, let's move forward and create our first commit with this command:

```
$ git commit -m "First commit"
```

We will see results like this:

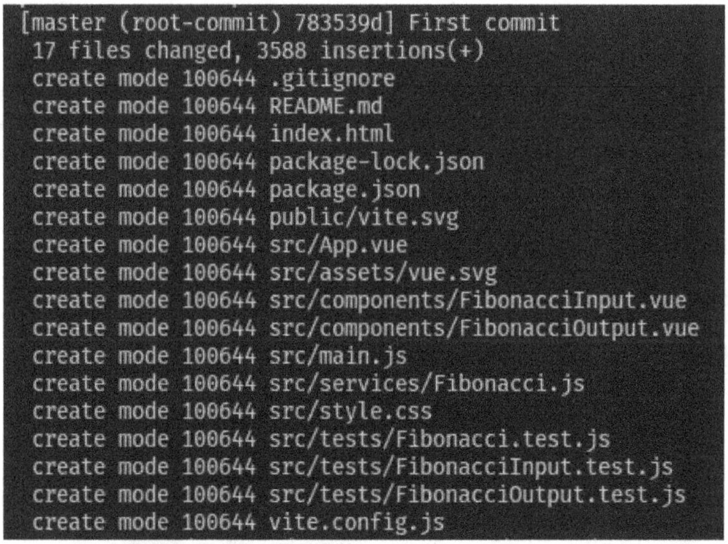

```
[master (root-commit) 783539d] First commit
 17 files changed, 3588 insertions(+)
 create mode 100644 .gitignore
 create mode 100644 README.md
 create mode 100644 index.html
 create mode 100644 package-lock.json
 create mode 100644 package.json
 create mode 100644 public/vite.svg
 create mode 100644 src/App.vue
 create mode 100644 src/assets/vue.svg
 create mode 100644 src/components/FibonacciInput.vue
 create mode 100644 src/components/FibonacciOutput.vue
 create mode 100644 src/main.js
 create mode 100644 src/services/Fibonacci.js
 create mode 100644 src/style.css
 create mode 100644 src/tests/Fibonacci.test.js
 create mode 100644 src/tests/FibonacciInput.test.js
 create mode 100644 src/tests/FibonacciOutput.test.js
 create mode 100644 vite.config.js
```

Figure 9.10 – Results from the first commit

With these simple lines, we have started to keep track of our source code. Now, as mentioned before, we have committed our initial code to the `master` branch. Git allows us to make an instant copy of the state of our code, much like a screenshot, and continue working from there without affecting the original. This is called **branching** and is an important part of using Git.

Managing branches and merges

Using branches to control our development is a very good way to move forward on a certain footing. Here are the most common commands to manage branches:

Action	Command example
Create a branch and switch to it	`$ git checkout -b [branch_name]`
Create a branch but stay in current	`$ git branch [branch_name]`
Delete branch	`$ git branch -d [branch_name]`
Switch to a branch	`$ git checkout [branch_name]`
Merge a branch with current	`$ git merge [branch_name]`
Check current branch	`$ git branch`

Once you have moved to another branch, you can perform all the regular Git operations (edit and delete files, etc.) without affecting the other ones.

Merging conflicts

When merging multiple branches together or with `master`, it is possible and more than likely that some files will have a discrepancy with the current branch. In this case, the *merge* will fail, and the user will be prompted to solve the differences. What Git does is mark the target file (the file in the current branch) with markers in the text for the user to edit. Once these have been edited, the file can be staged and committed, thus ending the *merge*. Let's try that without code, by purposely creating a discrepancy to fix. Follow these steps:

1. Create a new branch, dev, with this command: `$ git checkout -b dev`.

2. Edit `index.html`, adding in line 11 (before the script tag) the following: `<div>A div created in branch dev</div>`.

3. Save the file, stage it, and commit the change with `$ git add index.html` and then `$ git commit -m "added div in dev"`.

4. Now, we will move to the `master` branch with `$ git checkout master`.

5. Notice how line 11 with the div has disappeared from `index.html`. This is because that edition was never made on this file. Now, add in that line the following: `<p>This change was made in master</p>`.

6. Save the file, stage it, and commit it with a different message (look at *step 3*).

    Now, we will try to merge both branches and, as `index.html` has been committed to both with a different code, it will fail! To start the *merge*, run `$ git merge dev`.

    You should see an error on the terminal, and new lines added to `index.html` indicating the discrepancies. In our code example, it looks like this:

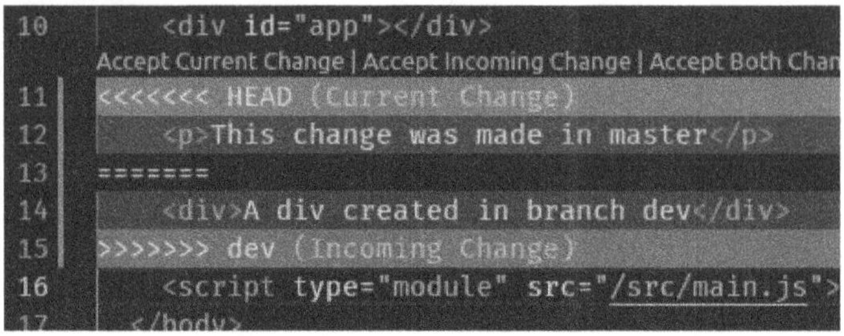

Figure 9.11 – Merge conflict

7. To resolve the conflict, just edit the source code to your best judgment (also delete the extra labels added by Git), then save the file, stage it, and finally make a commit. You will receive a message indicating that the *merge* has been resolved.

Working with branches and resolving mergers when they appear is a common practice and quite useful, but we are still not using the full potential of Git. As you remember, Git is a distributed SCMS, and this relies on its great potential. Enter the remote repository...

Working with remote repositories

In the same way that we work with a local repository, Git can also synchronize code with a remote repository. This enables team members from anywhere in the world to collaborate together in the same code base, resolve conflicts, and also synchronize their own code with others' contributions. Working with a remote repository involves the following steps to set up:

1. The remote repository must be created, and a URL provided to connect to it.

2. We add the remote repository as a new origin to our local repository with the following command:

```
$ git remote add origin URL
```

3. We set our master branch to synchronize with the remote repository:

```
$ git push -set-upstream origin master
```

4. We retrieve changes from the remote repository:

```
$ git pull origin master
```

5. We submit our changes to the remote repository:

```
$ git push origin master
```

Once you have done *steps 1* to *3*, the regular activity will involve *steps 4* and *5*. These activities will keep your local repository synchronized with your remote repository. In practice, modern IDEs such as Visual Studio Code will already provide you with graphical tools to make these operations, and this results in more convenience when you are working on a project. They also include visual tools to resolve conflicts during *merges*.

Setting up a Git server for your local network is outside the scope of this book, but this introduction would not be complete without a word about **GitHub** and **GitLab**. It is common that when people first hear about Git, they associate it with GitHub, which is understandable, as the latter has a much more popular media presence. GitHub is not Git. It is a web platform that provides tools built on top of Git, to host online projects hosting remote repositories. Thus, you can perfectly work with Git locally, and synchronize with a GitHub or GitLab remote repository. This is the most common case.

GitHub provides messaging and documentation tools, and much more – even additional services that allow detecting events in our repository to trigger certain actions and services, some provided locally (at a cost), others remotely (for example, webhooks). For example, it is possible for you to commit locally, push the changes to the `master` branch on GitHub, and have a whole set of routines started, from compilation to website presentation. Again, managing all these options is outside our current scope in this chapter, but the important thing to remember is that all of this is based and built upon Git, so if you understand *how* it the works and *what* it does, you have a solid foundation to move ahead with other tools and services. There is one more concept that has become familiar with this topic, Continuous Integration and Delivery, which we will see next.

Continuous Integration and Delivery

Continuous Integration (CI) is a practice enabled by the technologies we have seen thus far, where developers commit their changes to a central (remote) repository as frequently as possible. The central repository detects the incoming changes and triggers automated tests against the code. Then, it compiles/builds the final product. This is done continuously, as opposed to the practice of merging and compiling on a given date before launch.

Continuous Delivery (CD) builds on top of CI, by also deploying the released product to its final location. You can configure this process to create preliminary versions of software or web applications, (for example, betas, nightly builds, etc.), and program a release date for the end location and delivery to the customers (sometimes, this last part may involve a process of its own and is called **Continuous Deployment**). Both services mentioned before (GitHub and GitLab) offer these types of services.

By using these concepts, it is possible to set up a whole automated workflow from your desktop to the web, where a simple Git commit and a push to the server would trigger your application to be tested and published in its destination online. The way to implement this workflow is particular to the tool used to implement CI and CD.

Summary

In this chapter, we have covered very important concepts regarding the care and quality of our code. We have learned how to install official tools to perform automated tests in our code and components, as well as how to keep track of changes and management in the source code. While the examples and information provided here are introductory, they are detailed enough to implement them in your own projects and keep your learning skills growing. The concepts of CI and CD, as well as services provided by online repositories, also give you a solid foundation to learn to use them, as they all are based on the functionality provided by Git. All these tools have professional value for a developer and are required in the industry today.

Review questions

- Why is automated testing important? Does it eliminate the need to perform manual testing?

- What is necessary to test our Single File Components in Vue?

- What is source control, and why is it necessary?

- What is Git, and how is it different from GitHub/GitLab?

- When you modify a file in a branch, does that modify it in all other branches? Why does or doesn't this happen?

- Are the commands to control Git the same across all platforms?

- What do CI and CD stand for, and what is the value they add to a workflow?

10

Deploying Your Application

Working on and developing our application would come to a sad end if we could not publish the final product. While quite straightforward, presenting our application on the internet does require attention to a few details and being familiar with some terms and hosting possibilities.

In this chapter, we will learn about the following:

- What is involved in publishing a web application on the internet

- Considerations for building our application for deployment

- Becoming familiar with the terms and processes to register a domain

- Configuring a web server to host our **Single-Page Application (SPA)** or **Progressive Web Application (PWA)**

- Securing our web application's server with Let's Encrypt

The main purpose of this chapter is to give you the tools to understand the steps needed to publish and secure a website, and by extension, also our SPA or PWA.

Technical requirements

This chapter is mostly informative, but a few configuration files have been uploaded to the book's repository as examples, which can be found here: `https://github.com/PacktPublishing/Vue.js-3-Design-Patterns-and-Best-Practices/tree/main/Chapter10`.

What is involved in publishing a web application?

Publishing a Vue 3 web application is not that much different from doing so for any other website, save for a few key differences. In this chapter, we will consider a clean installation, meaning that we will procure all the elements involved by ourselves. At its most basic, we need to consider the following:

- A domain name for our site/application

- The destination path for our application

- The hosting service or type

- The web server software

- Procuring a security certificate

The preceding items also give us a simple formula for our preparations. Let's go one by one, explaining each necessary term and concern as we move forward.

Domains, Domain Name Servers (DNS), and DNS records

Every computer connected to a network receives a unique identificatory address to distinguish it from the others on the same network. These are called **Internet Protocol (IP)** addresses, and nowadays, there are two in operation – IP versions 4 and 6.

- **IPv4**: Addresses are comprised of four numbers separated by a dot. The number range is from 0 to 255. There are a few reserved addresses with a special meaning, such as 127.0.0.1, which represents a loopback to our own computer. These addresses can also have a mask that defines a sub-segment in a network. Most likely, your home network uses this protocol internally.

- **IPv6**: This provides a significantly larger address space, with eight groups of four hexadecimal digits, separated by a semicolon. Being so large, the protocol also provides ways to simplify the notation by removing leading zeros and replacing all zero segments with an empty segment. For example, the loopback address equivalent to IPv4's 127.0.0.1 in IPv6 is 0000:0000:000 0:0000:0000:0000:0000:0001, which then can be abridged as 0:0:0:0:0:0:0:1 or just ::1.

There is much more regarding network addresses, but only with this brief introduction, it is already possible to see a usability issue here. These addresses work great for computers but don't play nice with "*human memory.*" On the great internet, with millions of connected computers, using only IP addresses for navigation would be impossible. That is why there are special servers in the infrastructure of the web that provide a conversion from a "*human-friendly name*" into the right IP address. These friendly names are called **domains**, and the servers that provide the conversion are the **Domain Name Service (DNS)**. All this is regulated by the **Internet Corporation for Assigned Names and Numbers (ICANN)** organization.

Domain names are what we use every day to access any website or application on the internet today. These are bought for a limited time from an entity authorized to sell them, called a **registrar**. Once the period has expired, there is the possibility for a limited time to renew the domain, and if not, it can be acquired by anyone else. Usually, domains are sold in terms of years, and prices vary greatly from a few cents to thousands of dollars. Domains are also organized in groups, separated by dots from right to left, as shown here:

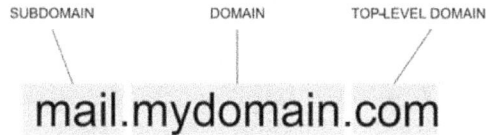

Figure 10.1 – The parts that make up a full domain name

The top-level domain is managed by the ICANN, and while `.com` for commercial websites is the most famous, there are plenty of others to choose from, such as the following:

- `.org`: For organizations

- `.net`: For networks or portals of corporate intranets and other organizations

- `.mil`: For military use

- `.gov`: For official government sites

New top-level domains are created often. You can find a growing list of them here: `https://en.wikipedia.org/wiki/List_of_Internet_top-level_domains`.

When we buy a domain (such as **mydomain** in the previous figure), it is attached to a top-level domain of our choosing. **Registrars** give us the option to select a domain and check whether they are available for purchase. To be of any use, the domain needs to be registered on a DNS to point to an IP address. The way to do that is to create **DNS records**, which is often done through the same registrar that sells the domain, or we can record in the registrar the DNS that will have the destination IP. There'll be more about this later, but for now, just keep present the concept in your mind. The most common DNS records for a domain are as follows:

Record Type (name)	Value and description
A	An IPv4 address. This is the main record that points to the public IP of your server.
AAAA	An IPv6 address. Points to the public IPv6 address of your server.
CNAME	Creates an alias to a domain, so you can point multiple domains to the same destination without creating multiple A/AAAA records. This can be used to create subdomains.
TXT	A plain text record, which is often used with some form of validation of ownership for a domain.

Table 10.1 – DNS record types

Depending on the registrar and the service hired, you may never see or have to deal with these records, as some registrars/web hostings manage them automatically for you.

Subdomains do not need to be bought from a registrar, only configured. You can create as many as you please for your own domain. Some common subdomains are as follows:

- www: For **World Wide Web**, or a web page. Nowadays, this subdomain is often used as a synonym for the domain.

- app: For applications.

- admin: For administration access.

- mail: For email services.

Using subdomains, you can host multiple websites off the same domain/host. We will see later how to configure one for our application on our web server. At this point, what we need to remember is that a domain or a subdomain will point as a final destination to your server.

> **A note about the loopback address**
>
> Following the previous examples, the "domain" name reserved for the loopback (home) address is localhost.

The domain where our application will be hosted is the first step to having a presence on the internet. With it, we are in place to move to the next consideration – where it will be placed in that domain.

Considerations for building our application for deployment

Once we have our domain/subdomain, we need to decide (or know) on which path the application will be located. The path is what follows the domain, in segments separated by a forward slash (/) – for example, mydomain.com/store/product.html. These sections are called "paths" because they follow the same directory structure as mirrored in the local storage. Our application will be served through a **web server**. These amazing pieces of software are built around the concept of serving files following the directory structure (folder/subfolder/file...). Internally, our server will match a domain request to the files in a local directory. Here is where we need to know whether the Vue application will be placed at the root, or on a path (directory), as we need to indicate this for the build process if we are using Vue Router in the web history mode (review *Chapter 5, Single-Page Applications*, if you need a refresher on this topic). In this case, we need to make two modifications:

- Indicate the *root* path for the application in our Router configuration

- Configure the web server to alter the directory/file service and route all the requests to the index.html file

If our application is placed in mydomain.com/**app** using the web history mode, then we need to alter our router definition by passing the "base path" to the creator function. So, if we look at the router from our SPA example application in *Chapter 5*, *Single-Page Applications*, we can modify it as follows:

/chapter 5/to-do SPA/src/router/index.js

```
import { createWebHistory } from "vue-router"
// ...
router = createRouter({
history: createWebHistory('/app'),
routes,
// ...
}
})
```

Note the minor change, where we pass the base path to the createWebHistory constructor instead of using createWebHashHistory(). Of course, if the application uses the hash mode, it doesn't matter where it is placed in our path. This is because in this mode, all routing navigation will be passed after the hash while pointing to our index.html file. For example, if our router has a route called / description, using the hash mode will make the address mydomain.com/app**#description** (the hash mode) instead of mydomain.com/app/**description** (the web history mode).

> **Hashes in web addresses**
>
> The hash in an address indicates a link to a section on the page/file, according to the HTML standard, and is used by Vue to manage the defined routes when in the hash mode.

Having provisioned the destination of our application, we can now build the production code through Vite with the following command on the terminal:

```
$ npm run build
```

By default, the final production-ready files for our entire application will be placed in the /dist folder (at the same level as our /src folder). Now, with our built distribution files, we are ready to upload them to a server once we have set the proper configuration.

Web server options and configurations

When is time to upload our application to a server, we are faced with many options, based on the type of service and web server application provided. This combination of items is usually referred to as the "hosting" server, which includes the operating system, machine configuration, architecture type, and especially, the web server software. Here is a list of some of the most common options for each category:

Operating system	Linux or Windows	For our Vue 3 application, this choice is inconsequential
Hosting type	Shared	Our application will reside in a server in a private area of the storage but will share all the resources with other applications. Access to configuration is usually done through a web control panel
	Virtual Private Server (VPS)	We are provided with a virtual machine with full access to the entire configuration and resources, usually through a direct connection with a remote terminal
	Managed VPS	Like a VPS, but we are provided with a web control panel or other services to manage the machine
	Private server	Here, we rent real hardware from the hosting provider and have full liberty to all their resources
	Self-hosted	We connect a server directly to the internet by our own means and an internet connection
	Collocated	We provide a server to a server farm, where they take care of physical needs. We manage the server remotely with full control of software and hardware
Web server	Apache HTTP	This server is stable and heavily used in Linux and shared hosting
	Nginx	A smaller and fast web server, famous for managing very well a large number of concurrent connections, with efficient use of resources. Very easy to manage and very popular for VPS hosting

Table 10.2 - Common hosting options per category

In the case of Vue 3 applications, we aim to have a web server that is fast and reliable to attend to multiple requests simultaneously by serving static files. We do not require much CPU processing power, as we do not run code on the server, so our requirements for hardware and software are very low, so much so that practically any "static file server" would do. Most likely, our application will be part of a much larger infrastructure with other requirements, but those used to serve our Vue 3 app in and of itself are low.

The key consideration here is, again, whether we are using the web history mode in our router. In that case, we need to include a configuration to the web server software to direct all requests to the entry point of our Vue application (our `index.html`) when a request does not match the standard (a file in the folder directory). This may sound complicated but is rather simple. Directly from the official Vue Router documentation, here are examples of the two web servers.

Apache HTTP Server configuration

The Apache HTTP Server is used heavily in shared hosting providers and allows us to alter the configuration for requests by placing a single file in the root directory of the web application. This is very convenient and simple, but it does require that the hosting provider has enabled (or through the administration panel, allowed the user to enable) a special module that allows us to rewrite the incoming requests. The official documentation (`https://router.vuejs.org/guide/essentials/history-mode.html#apache`) shows this example:

/.htaccess

```
<IfModule mod_negotiation.c>
  Options -MultiViews
</IfModule>
<IfModule mod_rewrite.c>
  RewriteEngine On
  RewriteBase /                                    //1
  RewriteRule ^index\.html$ - [L]
  RewriteCond %{REQUEST_FILENAME} !-f
  RewriteCond %{REQUEST_FILENAME} !-d
  RewriteRule . /index.html [L]
</IfModule>
```

The preceding file should be placed alongside our `index.html` file. Every incoming request will then be routed to it and captured by Vue Router in the web history mode. Also, note in line `//1`, the `RewriteBase` rule. Here is where we change the path of our application, if not placed at the root of the domain.

Nginx server configuration

In the case of VPSes and private servers, the NGINX server is quite popular for how flexible and performant it is. It can behave as a reverse proxy, load balancer, and much more. Installing this server in a VPS with Linux/Windows is rather trivial, but we will not cover it here. You can see the documentation for each system at `https://www.nginx.com/resources/wiki/start/topics/tutorials/install/`.

Unlike Apache with the `.htaccess` files, we need to modify the server configuration file for our site. In Linux, this is usually placed in `/etc/nginx/sites-available`. The file follows a simple schema where, for each virtual server, we declare the location path (as in the domain path) and the location on the local storage (the directory or folder). Here is an example file from a Linux server:

/etc/nginx/sites-available/default

```
server {
```

```
    listen 80;
    index index.html;
    root /home/user/www;                              //1
    server_name www.mydomain.com mydomain.com;        //2
    location / {                                       //3
        try_files $uri $uri/ /index.html;             //4
    }
}
```

Let's look at the preceding code:

- In line //1, we place the local storage absolute path to our application.

- In line //2, we declare the domains and subdomains that will be associated with this server block.

- In line //3, we declare the location path to process. In this example, we are placing the application at the root (/). If placed in mydomain.com/app, we would write location /app.

- Finally, in line //4, we tell NGINX to try to find a valid directory/file and, if not, pass it through to our index.html file.

As before, if we are using the web hash mode, then we do not need to do these changes. We can just use the default configuration to serve the files from the disk.

Other servers

There are many other servers in use and possible configurations that are not possible to cover here. However, the official Vue Router documentation has very good examples for other servers and guidelines for those not covered. You can find the reference at this link: https://router.vuejs.org/guide/essentials/history-mode.html#example-server-configurations.

Let's see now how to move our files onto our online server.

Transferring your files to the server

Armed now with the domain pointing to our server and the configurations in place, it is time to upload our distribution files. Depending on your choice of hosting, this could be done through a web interface, a **File Transfer Protocol** (**FTP**) application, or secure transfer over the **Secure Shell Protocol** (**SSH**). For the last two options, it is recommended to use an application that takes care of the heavy lifting. An excellent option is to use FileZilla (https://filezilla-project.org/), which handles the aforementioned protocols. It is available for Linux, Windows, and macOS.

As we mentioned in *Chapter 9, Testing and Source Control*, you can also configure your VPS/server to pull the source code from a remote repository using **Git**. In this case, the application folder will be configured to point to the /dist folder in the local repository. We could, for example, open a remote terminal to the server, trigger a synchronization (pull), and then bundle the application on the server itself, pull a branch that already has the application bundled, push our commits directly to the server, and so on. There are many options when using Git, and a few more when using a service such as GitHub or GitLab with powerful tools for *continuous integration and delivery*. This is a topic worth exploring if you do not want to use S/FTP applications or want to automate the process. Each implementation would be specific and out of the scope of this book, so we will move on to the next topic, assuming that our files are now on the server.

Protecting your web application with Let's Encrypt

Internet addresses are included at the very beginning of the protocol being used. By default, all web navigation is done using the **Hypertext Transfer Protocol** (**HTTP**), which, while foundational, is not considered secure. When an encryption layer between the client and the server is provided, then the communication is done over **HTTPS** (the **S** standing for **Secure**). This encryption layer is provided and validated by a certifying authority, so the certificate must be bought from such one. Hosting providers usually have the option to buy and install one on their servers, but there is also a free and reliable alternative provided by the **Let's Encrypt** foundation (https://letsencrypt.org/).

To install a *Let's Encrypt* certificate, you need shell access to your server. If not, then you must rely on the service provided by the hosting. The list of certified hosting providers compatible is here: https://certbot.eff.org/hosting_providers.

In the case that we have access to a server through remote shell access, the process is also straightforward. The Let's Encrypt foundation and the **Electronic Frontier Foundation** (**EFF**) provide an application called **certbot** (certification robot), which automates the installation of security certificates and also configures the local web server files to use HTTPS. In this case, we have two options:

- Install a certificate for each domain and each subdomain

- Install a *wild card certificate*, which covers each domain and all possible subdomains

The instructions to install the *certbot* and then run the process differ for each operating system and web server, and the type of certificate mentioned. Because of this, the EFF provides a web page with configurable options for each possible variation and easy-to-follow steps. The wizard can be found here: https://certbot.eff.org/.

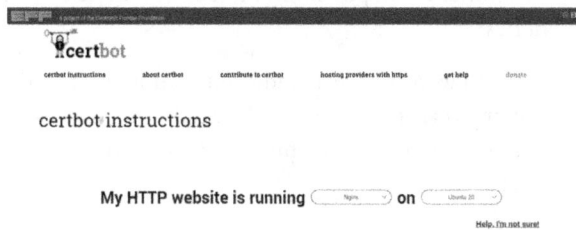

Figure 10.2 – Certbot instructions for NGINX and Ubuntu 20

In general, the instructions follow these steps:

1. Install **certbot**.

2. Run **certbot**. This will present a series of options, with all the found dominions found on the given web server.

3. Select the type of certificate to install.

4. Select, if active, the auto-renewal of the certificate. Rejecting this will require manual renewal.

The free certificate is only valid for 3 months at a time, as opposed to a commercial certificate, which can be bought for more time. After 3 months, it must be manually renewed. Luckily, *certbot* includes a function to perform automatic updates before the expiration period is due.

Even for simple test applications, it is important and recommendable to always protect a website with a security certificate. Let's also remember that having a security certificate and serving an application over HTTPS is a hard requirement for PWAs.

Summary

In this chapter, we covered the basics for publishing our Vue application in its own public space on the internet. We also learned important concepts for understanding instructions when buying and reserving a domain and setting DNS records if and when prompted to do so. We also learned about how to accommodate our bundle configuration when using the HTML5 history mode in Vue Router, the different types of online hosting we can hire, options for copying our distribution files onto the production server, and guidelines for securing our website with a free Let's Encrypt certificate to serve our applications over HTTPS. All of these are important skills, and you will benefit from having the experience of executing these skills at least once.

With the deployment of our application, we have covered in this book the main steps and topics to construct a Vue 3 application, from the introduction of the framework all the way to testing our individual components and installing our production-ready files in a web server. In some cases, we have gone beyond the basics to see advanced topics, which are an important addition to a professional developer. If you have followed the concepts and code examples this far, you have acquired important skills to help you in your professional development. But this is not the end of this book, as you can find additional bonus content next.

I extend my appreciation and gratitude to you, dear reader, for your interest in Vue 3 and in acquiring this book, which summarizes many years of experience developing applications. I hope it becomes a source of reference and encouragement to keep learning, and I wish you the best of success in your personal and professional career.

Sincerely,

Pablo D. Garaguso

Review questions

- What is a top-level domain, and how is it different from a domain?

- Is there a limit to the number of subdomains we can create for our domain? Why?

- What is a DNS? What are DNS records?

- What are the options currently for publishing your web application on the internet?

- What considerations in our code and web server do we need to make when using Vue Router in web history mode?

- What are some common web servers found in shared hosting and VPSes?

- What is certbot and what does it do?

Bonus Chapter - UX Patterns

A long time ago, computer software was sold with heavy, printed manuals and in many cases, also with specialized training. While some of this still applies to corporate applications and specialized hardware, such a concept has almost completely disappeared from the consumer market. Imagine that, for each website, you had to read a manual and attend three months of training classes before even using it. How come, today, a user can be presented with a new application and somehow manage to accomplish basic tasks at first sight? No doubt, a great contributor to this current situation is the years of study and progress in the areas of human-computer interactions, a discipline almost as old as computers themselves.

In this chapter, we will take a step back from the intricacies of inner-working software patterns and architecture and look at some of the well-established patterns for designing a **user interface (UI)** and **user experience (UX)**. We will do the following:

- Understand the differences between UI and UX
- Become familiar with common UX patterns in web applications
- Learn about the evil side of UX with dark patterns.

By the end of this chapter, you will have a broad understanding of what UX patterns are, the principles they are based on, what the most common approaches are today, and the effects they produce for the user, be they positive or negative. You will also learn and develop a common language to interact with UI and UX designers and other developers.

Technical requirements

This chapter is mostly informative; however, small examples are discussed and implemented in Vue 3, as they demonstrate the use of a technique not seen this far in the book. For the complete source code, please refer to the project folder for the chapter in the book's repository at https://github.com/PacktPublishing/Vue.js-3-Design-Patterns-and-Best-Practices/tree/main/Chapter11.

Check out the following video to see the Code in Action: https://packt.link/5ymkr

UI design versus UX design

It is common to hear these two terms associated or used indistinctively, and in some cases, both responsibilities are merged into the same role in a team, which adds to the confusion. While there is some overlapping, as often happens in computer science disciplines, we will focus on the differences for learning purposes:

- **UI design** is in charge of defining the visual language to represent information and capture user input to interact with the application. It covers the design of visual styles, typography, interactions, colors, sizes, animations, sounds, and so on that make up the interface between humans and computers (software and hardware). This applies to visual mediums (web, mobile, etc.) as well as other types such as natural language interfaces (think of AI assistants such as Siri, Alexa, etc.).

- **UX design** includes the conditions that affect and guide the UI design but encompasses a broader view focused on the user's perception of the system, company, or feature they come to interact with. It involves elements beyond software or hardware, with aspects such as support, promotion, post-services, and so on. The aim is to create a wide and, hopefully, successful positive experience from the user's perspective. Under this definition, it also collaborates with other disciplines, such as marketing, customer support, distribution, product management, brand recognition, and so on. It has the primary objective to alter or create the user's perception that the product, service, or system is easy to use, efficient, and above all, useful for their purposes.

  Understanding the principles and objectives of each discipline will help us develop better software and have a common ground for understanding when collaborating with those in these assignments. Without UI and UX, even the best-produced software may fall into oblivion. Software history is full of examples of companies that went the way of the dodo bird, even with superior products to their competitors, for neglecting the user experience or having a bad visual design. Sometimes, the cover of the book is as important as the content...

The principles of UI design

At its core, UI design has the objective of creating an interface that the user finds easy to use, efficient, informative, and enjoyable. Marketing concepts such as user retention and satisfaction rely heavily on product design. For our purposes, we will limit the introduction of UIs to on-display application interfaces (presented through a visual medium, such as a screen or touch device).

There is ample documentation studying UI design in detail, with engineering precision and well-defined industry standards. Each aspect has its own set of rules that a good UI designer needs to keep in mind. A web designer will see things differently from an industrial designer, for example. In our case, most of the patterns for UI design have been included since the beginning in the HTML standard, so most, if not all, of what we will see as patterns are already familiar to you and the end user. How they work, or the principles they follow, however, is not something that is commonly discussed or self-evident. For example, why is the "X" to close a window in the top-right corner? What does each different menu icon mean? Why does the **Start** button appear on the top or bottom-left corner of the screen? Why are some features hard to find, while others are accessible at first sight? All these questions have a solution rooted in UI design and UX patterns. With that in mind, let's review some of the UI principles, and then move on to the UX patterns.

Sufficient contrast or distinction between elements

This principle states that the elements of a page should be clearly distinguishable from each other and representative of their function. It reflects the need to organize the visual elements in a hierarchy using size, colors, typography, margins, and white space in such a way that each function is clearly represented and distinct from others. The main objective is to direct the user's attention to the focal point of the interface. Let's see the Packt home page (`https://www.packtpub.com/`) as an example:

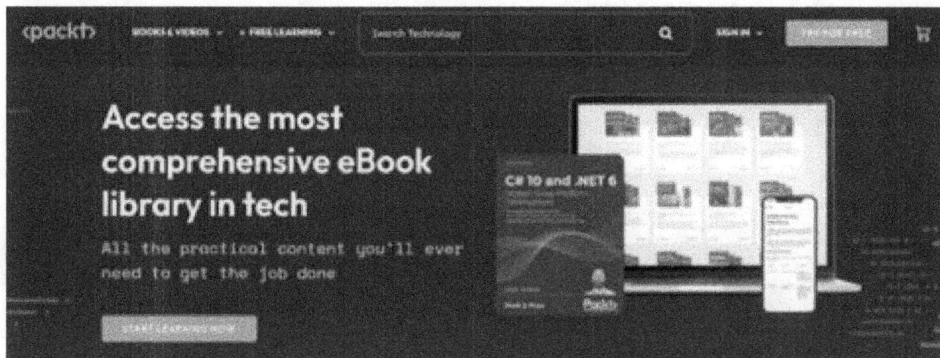

Figure 11.1 – Packt's home page and color contrast

In this example, most of the content uses a limited number of colors (a "palette"), and the two buttons have high contrast, which draws the attention of the user: the **TRY FOR FREE** and **START LEARNING NOW** buttons. Clearly, the designer has set the focus point for these two actions.

Related to this principle, there is a design "rule" applied in this screenshot that is handy to keep in mind: the *60-30-10 color rule*. This means that 60% of the section must have a base color (here, it is dark brown), 30% is a primary color (used for regular text, menus, and the image, which, here, is white), and 10% is reserved for the secondary color or high contrast (used for the focal point, or the "call to action" buttons in orange). Let's continue analyzing this page to see the other principles of UI design.

Stimulating repetition and being consistent

This principle is based on the concept that people learn through repetition. It indicates that the same task, even done through different interfaces, should emulate more or less the same repetitive behavior from the user. For example, if you ask a user to open a word processor, and you ask them to open a file, where will they try to click? Most "seasoned" users will take their mouse to the upper-left part of the window and look for either an icon that represents "**Open**," or the **File** menu. Why is this? It is because this placement has become standard, and we have learned through repetition where to find it. If you placed the **File** menu at the bottom right of the screen, most, if not all, of your users would have trouble finding it without direction.

Another example of repetition and consistency appears when displaying visual elements – more importantly, when they are part of a list or common set. Let's follow with an example from the Packt Publishing website:

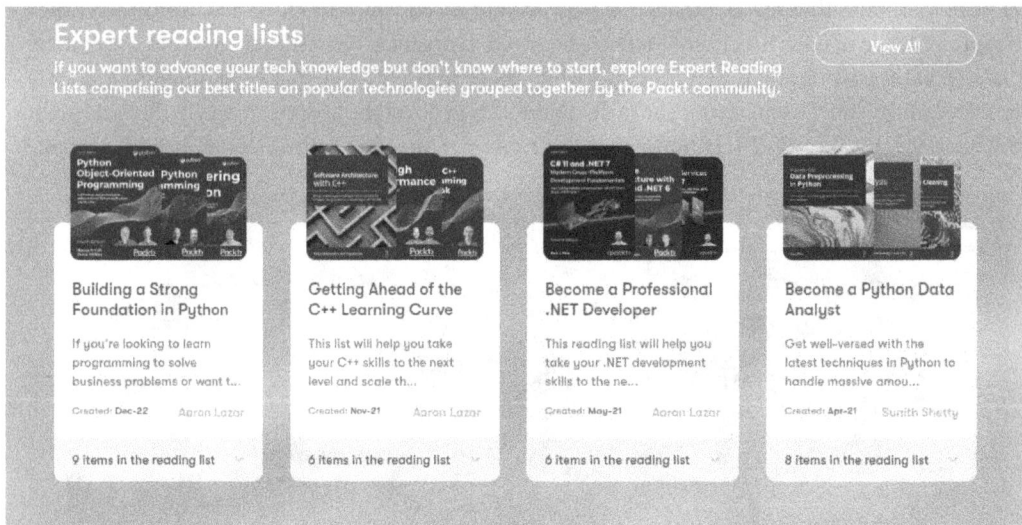

Figure 11.2 – Cards with books

In the preceding example, the designer used repetition to display the books in prime time, by using vertical "cards" for each item. Each card repeats a similar layout, colors, and format. Once you understand what one displays, the same applies to all the others: this is a repetition of visual design and is important, so the user doesn't have to "re-learn" the interface for each book.

In short, it is the repetition of placements, actions, and visual styles. Let's move to the next one.

Be mindful of alignment and direction

The alignment of elements (graphics, fonts, etc.) creates a sense of order and organization, showing that such elements belong to the same group or have the same weight or importance. We are mostly familiar with alignment (and spacing) when working with fonts, but the same concept applies to graphical elements such as icons, sections, images, and so on. From the previous figure, notice how the tabs are aligned, as well as the cards and the content therein. Just by alignment and style alone, we can distinguish what belongs to which natural group.

Another example of alignment is clearly seen when only using typography for menus and features. For example, in the footer of this page, even without the use of icons or visual boundaries, it is easy to see what option belongs to each natural group just by using space and alignment:

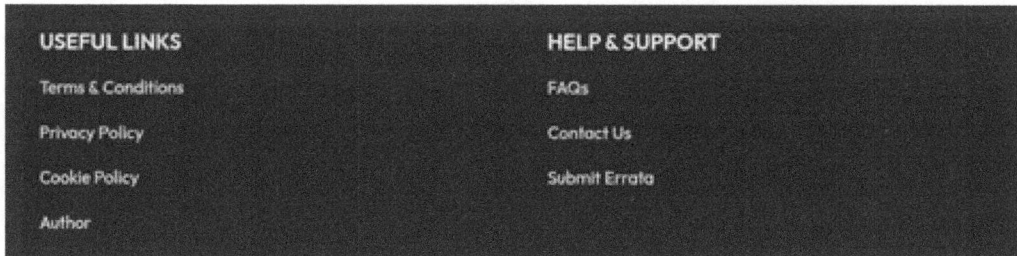

Figure 11.3 – Packt's website footer, using alignment to display natural groups

This example, though simple, already shows the use of contrast (bold versus normal font-weight), consistency through repetition, alignment, and also the next principle we will see here: proximity.

Use proximity and distance to show natural groups

This principle is simple to follow: place the elements that are naturally related close to each other. This makes it easier for the user to find and understand. Typography and iconography can also be used to show proximity and similar functionality. The famous "ribbon bar" introduced by Microsoft in its Office applications uses this concept heavily and soon became a standard. For example, here is a segment of the **Home** tab that deals with alignment, lists, spacing, and indentation:

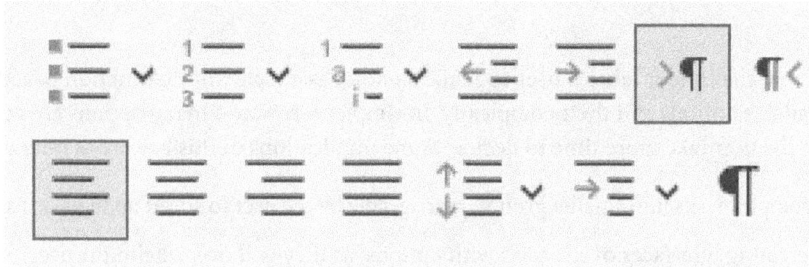

Figure 11.4 – Paragraph icons

Notice how the icons that modify the type of paragraph, spacing, and alignment are close to each other, without being "mixed."

These are not the only principles of UI design but are the most basic ones that we should keep in mind when developing our components. If you work with a team that includes a UI designer, you may receive templates or mockups, even storyboards, to implement into Vue components, as we saw in *Chapter 4, User Interface Composition with Components*. If instead, you are on a *"one-man band team"* and the design of the interfaces falls in your lap, these principles will help you greatly to create a professional and useful UI. But there is more…

Laws for UI design

Just like we have principles, several studies have issued or identified "laws" of design, which are measurable functions to predict certain software attributes, such as usability and friendliness. In particular, the following stand out.

Fitt's law

This law specifies that the time taken for a user to acquire a target is a function of the distance from the pointer and the size of the target. The important concepts here are the distance and size of the target: the longer the distance between targets, the larger these should be.

The application of this law is what places the window close button (**X**) in one of the corners of the screen (when the window is maximized), the **Start** button at the lower left of the screen, *Show desktop* at the bottom right of the screen, and so on. If a user moves the mouse in any of these directions, it will sooner or later arrive at these targets, and continuing in the same direction will not affect the result. In the language of UI design, these buttons are "infinite" as once the cursor has reached a corner by the edge of the screen, continuing scrolling in the same direction with the mouse will still hit the target.

Another implication of this law is that functions that are similar should also be placed in proximity to each other (such as the examples in *Figure 11.4*). You can find more information about this law on Wikipedia, at `https://en.wikipedia.org/wiki/Fitts's_law`.

Hick's law

This law says that the time it takes a user to make a choice is a logarithmic function based on the number of available options and their complexity. In simpler terms, too many options are confusing and will make the user take more time to decide. Some implications of this law are as follows:

- Break complex tasks into smaller groups, manageable by the user to speed up their decision time
- Avoid creating interfaces overloaded with options, as they will overwhelm the user
- If a function is time-sensitive, minimize the options to the bare minimum for the situation

In everyday software, we can see the application of this law in multiple places – for example, in "installers" for a particular software, where successive windows with options are presented to the user during or before starting the installation in a sequential manner, as opposed to a form to fill in. More information about this law can be found on Wikipedia: `https://en.wikipedia.org/wiki/Hick%27s_law`.

Ben Schneiderman's eight rules

In 1986, Professor Ben Schneiderman published his book titled *Designing the User Interface: Strategies for Effective Human-Computer Interaction*, where he stipulated eight rules for interface design. These rules are as relevant today as on the day they were created, so they are worth mentioning here:

1. Strive for consistency (in actions, steps, placements, etc.).
2. Enable the use of shortcuts for common tasks (be it using a keyboard or icons).
3. Offer informative feedback (especially when an error occurs).
4. Design dialogs with an end.
5. Offer simple error handling, so the user can take action quickly and avoid serious errors by the user. The classical implementation of rules #4 and #5 are "confirmation dialogs" before performing permanent actions, such as deleting content.
6. Permit easy reversal of actions (thanks to this, we have *Ctrl + Z!*).
7. Support the sense of control in the user. Nothing is worse for a user than to feel that the machine is "doing its own thing, out of control." If you ever sent an 800-page file to print by mistake, and it took you a dozen pages before you could actually cancel the operation... this is what this is about.
8. Reduce short-term memory. The user can only keep in their short-term memory a handful of items and tasks at a time, and too many elements on a screen (menu, and so on) create rejection. This principle also relates to Hick's law.

For practical applications and an introduction to the topic, these principles, rules, and laws should give us a sure footing.

More information about these rules and Ben Schneiderman can be found on Wikipedia (`https://en.wikipedia.org/wiki/Ben_Shneiderman`).

UX design principles

UX design also has its own objectives and principles that apply to patterns. Above all, the main objective of UX is to provide a good perception to the user, to create a bond with the brand or product, by carefully tailoring a flow of interactions. In this case, the solutions must be as follows:

- *Useful and usable:* First and foremost, the application must do what it intended to do, do it well, and be easy for the user to use

- *Learnable and memorable:* The user must be able to learn and understand what information is presented and internalize it for future use as well

- *Credible and give control to the user:* When the user interacts with the application, it must feel that it is doing what the user intends and that the result is "safe" for the user

This last principle is very important. If the user feels that the application somewhere during the interaction has "lost control" of what is going on, it is a recipe for disaster. Sadly, this occurs all too often with the dark patterns that we will see later in this chapter, but now, let's see some good patterns that result in a good experience for the user.

Common patterns for data manipulation

These patterns are often matched by pure HTML elements, while others have emerged in recent years through the clever use of styling of such elements. These have become standard in the industry and are well understood by users at first sight. What follows here is a non-exclusive list with a short description of when to use each one.

HTML input elements

The standard input fields provided by HTML are a clear pattern for receiving input from the user. Nowadays, the input element has many variants due to the `type` attribute, allowing for input other than plain text. Used in forms and validation libraries, these elements are as good as they come, ready to read and format from text and numbers to URL, date, time, images, and colors. The full list of available types supported by today's browsers can be found here: `https://developer.mozilla.org/en-US/docs/Web/HTML/Element/Input#input_types`.

For the most part, these elements are used with basic functionality and some heavy CSS styling. Use `inputs` (and `textareas`) any time that you need the user to enter text information. Browsers today offer native-looking widgets for the more complex types, such as date and color pickers.

Checkboxes, radios, and toggle switches

Checkboxes and radio buttons are provided natively by HTML and are presented to the user following the format of the local OS or environment. **Checkboxes** represent multiple options to the user, which they can select freely from the group. In contrast, **radio** buttons only allow the selection of one option from the list:

Figure 11.5 – Checkboxes on the left, and radio buttons on the right

With the release of the original iPhone, a new variant of the checkbox became very popular: the **toggle switch**. It is not provided natively by the HTML standard, but it is easily styled through CSS to "disguise" a checkbox. A toggle switch has two states, enabled and disabled, and it is often used to activate or deactivate a function or feature. This is an important distinction as the checkbox should focus on options or alternatives. Here is an example:

OFF   ON

Figure 11.6 – Toggle switches in each state

The toggle switch is in an "off" state (or false) when the toggle is to the left, and "on" when to the right. Often, switching also affects the color, showing it in a mute scale of gray when off or in vibrant colors when activated. Internally, these two states are often represented by `true` (on) and `false` (off), and they should be used to activate or deactivate settings, features, and so on. You can find the `Toggle` component that implements the styling and the `v-model` code in the repository for this chapter.

Chips, pills, or tags

This pattern encompasses short text (or "copy" in UI lingo) inside a round box. The copy can be accompanied by an icon to emphasize the state when selected, or an action such as an "**X**" symbol to remove it. Usually, pills are used in lists to display attributes, tags, categories, or other types of qualificatives for a given item. In some cases, they can be used also to select (toggle) items from the list. In this case, they rival or behave just like a checkbox but in a more visually pleasing manner. The implementation of chips is rather trivial and can be done with plain CSS in a single HTML element (for example, a span element).

Figure 11.7 – A list of items with pills

This is a short list of UI patterns to capture user input, but it does cover the most common types: the HTML inputs and styling variations.

Common patterns for data visualization

These patterns display information back to the user, either in response to a user action or an application event. What follows is a non-exclusive list of patterns.

Tooltips

This pattern shows the user some floating small text with information regarding the target element, usually when the user activates the element using some action (a hover, click, selection, etc.). The information is displayed above, below, or to the side of the element in the form of a "speech bubble" (like in comic books). Here is an example:

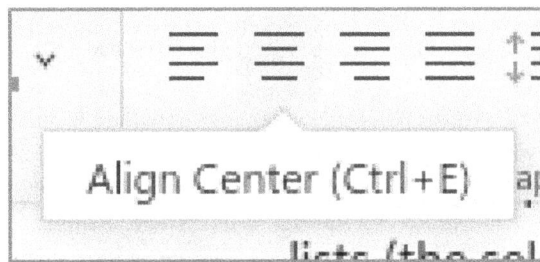

Figure 11.8 – A tooltip displaying the name/action of the icon and a shortcut

This pattern is mostly used to display help regarding the target object, but it can be used also to display contextual menus. For example, selecting a word from this paragraph in a text editor online will display a popup with a menu:

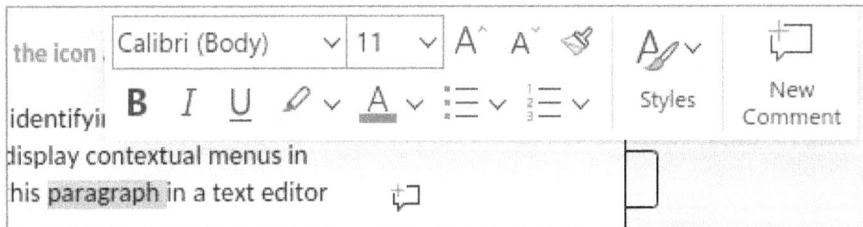

Figure 11.9 – A tooltip pattern used to display a contextual menu

There may be some discussion regarding whether this last use is a tooltip or a contextual menu, but the fact that it opens on selection would indicate the first. Traditionally, contextual menus are opened with a secondary action, such as the right mouse button (in Windows and Linux). In either case, the concept to remember for this pattern is to show information to the user on selection or pre-activation to help them decide what to do next.

Notification dots, bubbles, marks, or badges

This pattern consists of displaying a small icon on top of a larger icon to indicate that an event has happened and requires the attention of the user, yet it is not urgent. This small icon can be a dot, a bubble, and so on. If the notification has a number in it, it is also called a **badge.** Some examples of this pattern are as follows:

- The small circle with a number for new emails received in the email icon
- The double checkmark in a chat application indicating that the other party has received and read the message
- A small mark in a taskbar showing that an application has been opened

The key concept is to use some sort of small variance to the icon to indicate the need for future attention, but it is not urgent and won't affect the current activity for the user.

Toast notifications

This pattern is used in multiple applications and OSes. It consists of displaying, for a short amount of time, an overlayed floating window with quick information for the user. Often, it includes a short text fragment paired with a sensible icon. Depending on the OS, this can be displayed natively at the top center, right of the screen, or at the bottom right above the system tray. Web applications can either implement their own toast notification within the browser window or require permission from the user and display a "native toast notification" using the local OS. Here is an example of a toast notification:

Figure 11.10 – A native toast notification on the desktop

These notifications are useful to inform the user of changes in the environment that require their attention, to report back the result of an asynchronous operation (success, error, etc.), and so on. These calls to attention are ephemeral, so they should not be a critical part of an important workflow, except for the conditions just mentioned.

Carousel or image sliders

Carousel or image sliders are a pattern for displaying, in the same space, different sections with images and content in sequential order. Usually, using a timer, these sections are presented to the user a few seconds apart, also offering the option to skip to any of them through dotted navigation. While they were very popular some years ago, there are counterpoints to the use of these, as "impatient" users may never see the entire content of a carousel. In practice, the recommendation is to keep the list of sections as short as possible, somewhere between three to five elements.

Examples of this implementation abound on the internet, as they are implemented primarily on the landing page of a site. Practically all shopping and news websites make use of this pattern – for example, the Amazon landing page (www.amazon.de):

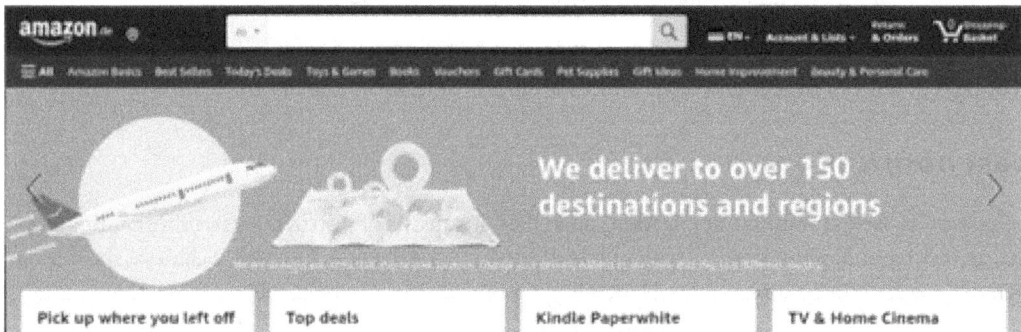

Figure 11.11 – Amazon's carousel displays offers from the store

Image sliders can be as big as a carousel but can also be much smaller and used to display thumbnails. Here is again an example from Amazon:

Figure 11.12 – Amazon makes good use of image sliders to display items to the user

While trivial to implement, the use of carousels and image sliders is a good way to display information to the user. There is a warning, though, that abusing this pattern could lead to an overwhelming and saturated experience, create confusion, and even trigger undesirable side effects, such as sensory overwhelming and content avoidance.

Progress bars and spinners

A progress bar is any element that gives the user an idea of how far down the workflow the current state of a process is. While the name seems to indicate a "bar," in practice, any element that displays progress through a limited number of actions falls into this category. The basic objective is to notify the user of the progress of tasks that take a long time and show that the system is "busy" working on them, thus providing visibility and a sense of control to the user. Progress bars are very important to prevent involuntary negative actions from the user. If a long-running task is being executed in the background (let's say in a web worker) without any feedback on the progress, the user may believe that the task has not started, has failed, or the computer has "hung up." It is a negative user experience to leave the user wondering what is going on. Here are some examples of progress bar elements:

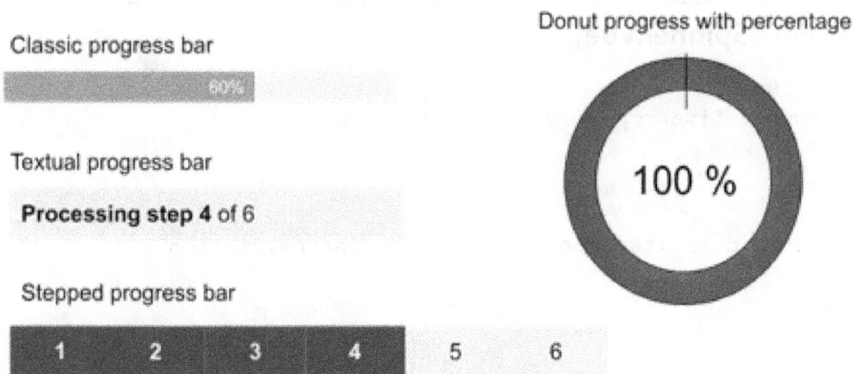

Figure 11.13 – Example of progress bars

In addition to styling, progress bars can also be used in an "undetermined state," meaning that it is not possible for the application to calculate how much time or how many steps a process may take; however, it still wants to inform the user that the system is busy and shouldn't be interrupted. The HTML standard does provide an element specifically for progress bars that can manage these situations (the progress element; see `https://developer.mozilla.org/en-US/docs/Web/HTML/Element/progress`), but there are other patterns to use in these cases, such as **spinners**.

As the name "spinner" indicates, this is an icon that "spins" on itself, giving the notion that the application is busy and working. Here is an example of a spinner with a text indicator:

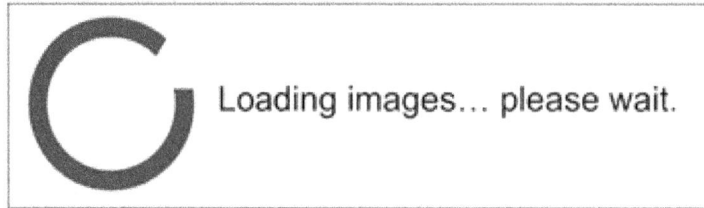

Figure 11.14 – Spinner circle, indicating the application is busy

This approach has been used in OSes and applications for some time now, so users understand its meaning. There is one caveat when using this pattern and that is that after a long period of time, it can create anxiety, so it is recommended to accompany it with some sort of indicator of action. Implementing a `spinner` component in Vue is rather trivial, and mostly CSS:

./components/Spinner.vue

```
<script setup>
const $props=defineProps(['caption'])
</script>
<template>
    <div>
        <span class="spinner"></span> {{ $props.caption }}
    </div>
</template>
<style scoped>
.spinner{
    display: inline-block;
    height: 1rem; width: 1rem;
    border: 2px solid;
    vertical-align: middle;
    border-radius: 50%;
    border-top-color: #06c9c9;
    animation: rotate 1s linear infinite;
}
@keyframes rotate {
    0%{ transform: rotate(0deg);}
    100%{transform: rotate(360deg);}
}
</style>
```

In this simple component, we only need to define a prop for the text, and a class for our spinning elements. The circle is made here by setting the border radius and defining a color for one border, so the spinning action is noticeable.

Pagination and infinite scroller

When we need to display a long list of items to the user, there are two patterns that come to mind as a well-understood solution: pagination and infinite scrollers.

In **pagination**, the dataset is divided into smaller parts of fixed size in sequential order. Each subset is called a **page** and is referenced by an ordinal (usually a number). This allows for easy navigation between pages, such as random and sequential access by the page number. Also, giving the same order function in the data allows a quick and easy way to "come back" to the data in different sessions. The element that allows the user to navigate in the paged data is usually referred to as a **pager**, and it is common practice to place it at the top and bottom of the item list. A typical pager might look like this:

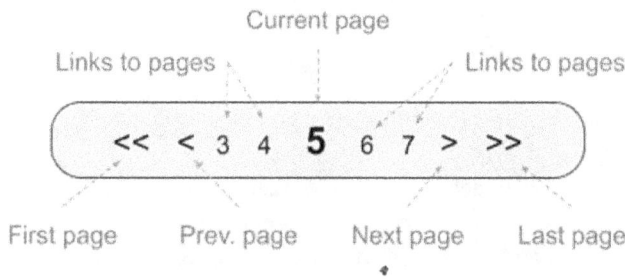

Figure 11.15 – The elements of a typical pager

In the preceding figure, you can see the different elements of a typical pager, usually used for tables or lists of content. However, this is not the only shape that a pager may take. It could, for example, use a drop-down menu for the page numbers, display ranges of pages, and so on. The important concept of this pattern is the division of the set and the quick navigation to each individual subgroup.

An alternative to pagination is to use an **infinite scroller**. In this pattern, the exact location of each item in the dataset may or may not be needed, and the items are presented to the user as they scroll the web page. When the user reaches the end of the list, new items are loaded on the page, chunk by chunk, until the user stops scrolling or the entire dataset has been loaded. Here is a graphical representation of this pattern:

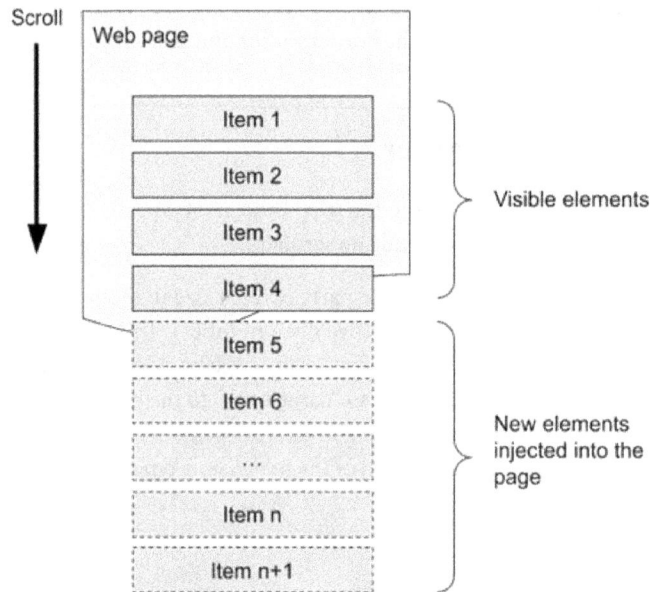

Figure 11.16 – The implementation of an infinite scroller

A number of techniques are used to detect user behavior and load or pre-load data. One of the easiest implementations is through the use of an **intersection observer**, a native element provided by browsers in JavaScript that triggers an event when such an observer intersects with other elements and, in this case, with the viewport. Since this is a new concept, we will implement a minimal infinite scroller that will look like this:

Figure 11.17 – The example infinite scroller limited to a `div` element

If you scroll down on the `div` boundary, by any available means (mouse, keyboard, etc.), the list will generate new items and will never stop, giving the impression that the scrolling is infinite. Here is the source code for that component:

./src/components/InfiniteScroller.vue

```
<script setup>
import { ref, onMounted } from "vue"
const _max_value = ref(30),                              #1
    _scroll_watcher = ref(null),                         #2
    observer = new IntersectionObserver(triggerEvent)    #3
onMounted(() => {
    observer.observe(_scroll_watcher.value)              #4
})
function triggerEvent() {_max_value.value += 20;}
</script>
<template>
    <div v-for="elem in _max_value" :key="elem">
        item {{ elem }}
    </div>
    <div     ref="_scroll_watcher"></div>                #5
</template>
```

The preceding component is minimal, but it does illustrate the technique. We will have a list of numbers with an initial limit that will trigger an overflow in the container (#1). This is important as, on the first load, the user will know that there is a scrollbar and more content below (as in *Figure 11.17*). Now, the technique here is to define a reactive variable, `_scroll_watcher`, pointing to `null`. This variable will later have the value of an element at the bottom of the list, which we mark as a reference (#5). We use `null` so Vue does not run any optimization at this point. In line #3, we create a new `IntersectionObserver`, and pass as value the reference to our `triggerEvent` function, where we will simply increment the boundary of our list. In our template, we use a `v-for` directive to generate the elements for each integer in the list, which should appear before our scroll watcher element. The magic happens in line #4, once our component has been created and mounted into the page. At this moment, Vue has already assigned to `_scroll_watcher` the reference for the HTML element, so we can pass it to our instance of `IntersectionObserver`. As we are using it with defaults, it will run the `triggerEvent` function each time the div in question appears in the viewport, which will happen when we reach the end of the list. In that function, we increment again the number of items, making Vue inject more elements into the web page and pushing the scroll watcher div again out of the viewport. This process repeats *ad infinitum*, giving us a simple but effective infinite scroller.

Beyond the implementation of the UX pattern, this technique is the correct approach to binding a reactive variable to an element in the DOM, and it saves us from writing direct JavaScript DOM manipulation, such as `document.getElementById("#someId")` and dealing with issues such as ID collisions. Vue resolves this for us.

Common patterns for interaction and navigation

These patterns control the interaction or offer the user options to control the process and navigation of the application. As usually happens, some of these patterns could also fit into other categories.

Placement of menus

The placement of menus is also a pattern that has been standardized in three basic layouts:

- **Menu bars** (horizontal) are often placed in a "sticky" position at the top of the screen (meaning they will not scroll with the page).

- **Navigation bars** (horizontal), mostly popular for mobile devices, are an iconography menu placed at the bottom of the screen to navigate to different sections of the application.

- **Sidebars** cover the full height of the screen, and with variable widths. These show menus with icons and/or text. In mobile and desktop applications where the screen's real state is important, there is usually an option to toggle it in or out of the viewport. The standard icon to trigger this feature has become the "hamburger" icon (see later in this chapter).

Following these standard placements for desktop and mobile will ensure that the user knows how to navigate the site easily. Often, some applications such as video games break these standards, but unless there is a powerful reason to do so, it should be avoided.

Breadcrumbs

Breadcrumbs are a hierarchical list of links that show the current position of the web page in the overall website organization. Each link allows the user to go back to a level, without having to use the browser's **Back** button or dig deep into the main menu. The current approach is to place breadcrumbs at the top of the page, before the main content. Here is an example of a breadcrumb path:

```
Home > Level 1 > Level 2 > Level 3 > Current page
```

The convention is to use the more-than symbol (>) to separate each navigation page, but there are many artistic licenses taken on this matter. Another representation for this navigation is to use a tree structure, such as this:

```
Home
    └── Level 1
      └── Level 2
        └── Level 3
          └── Current Page
```

This "folder-like" structure is not so common for main navigations but is mostly used for nested content such as comments and replies in forums.

Modal dialogs

A modal dialog is a small window that opens in front of the application, taking full control of the focus. It prevents the user from interacting with the rest of the application until the activity presented in the dialog is resolved. Modal dialogs should focus on one action only and provide sufficient information for the user to take action by giving clear options. Here is an example:

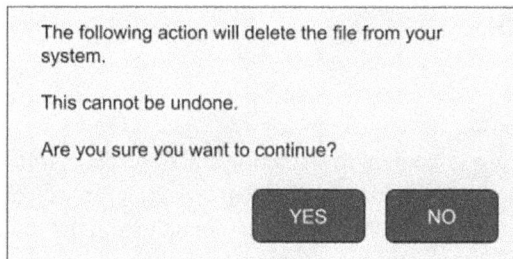

Figure 11.18 – A confirmation dialog

We have previously implemented a system to display modal dialogs in Vue 3, so you can review the code from *Chapter 5, Single-Page Applications*, to see an implemented approach.

Menu icons

Other than the plain copy (text) naming convention for menu items, there are a number of icons used today to show the user at first glance what kind of interaction to expect when presenting a menu. Here are some of the standard icons you can find and use today:

Figure 11.19 – Menu icons

And here are their descriptions:

- **Hamburger icon** (three horizontal lines): Reserved for *main menus* and navigation, this toggles the display of a wide site menu in a sidebar. These types of sidebars that show and hide on convenience are called **drawers** and are very popular in mobile sites and applications.

- **Kebab icon** (three vertical dots): Made popular by Google's Material Design, this indicates that there are more *options* for the current element or activity and that they will be displayed in a modal dialog.

- **Meatball icon** (three horizontal dots): This is displayed next to an item on a list and indicates that there is a pop-up menu with additional *actions* for the element.

- **Bento icon** (9 squares in a 3x3 grid): This is used to indicate a pop-up menu to navigate between different *applications* inside the same solution or environment.

- **Döner icon** (three stack lines with different sizes): This indicates options to *sort* the entries of a list using a selectable order option (usually in a popup). This icon, however, has not become as popular as the others.

Accordion menus

This pattern's basic use is to group content under a title and to display it only when the user selects it, allowing only one group to be displayed at a time. Nowadays, it is used commonly for *Frequently Asked Questions* pages, and in sidebar menus. Here is an example from the Packt website (`https://www.packtpub.com/`):

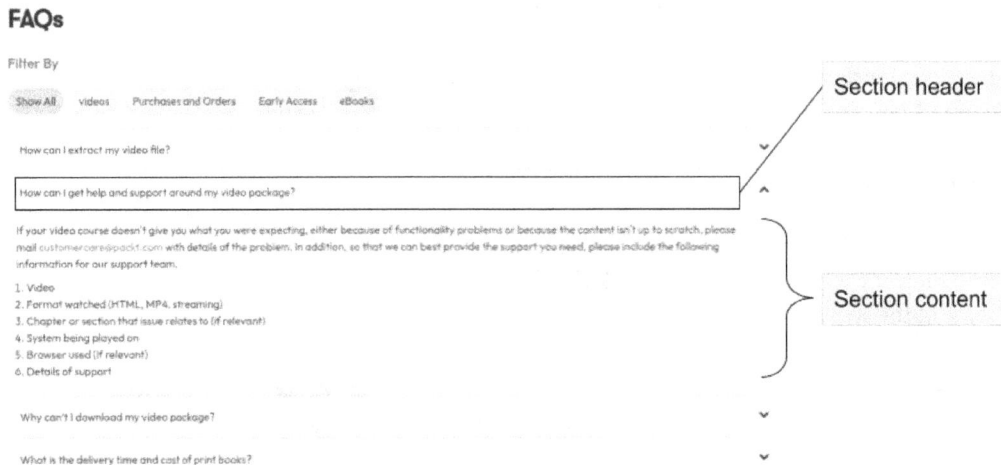

Figure 11.20 – Accordion menu used for FAQ

The accordion menu is a pattern well understood by users, and one rather simple to implement. It helps keep the design clean and allows the user to focus only on what matters to them.

Mega menus

Thus far, most of the patterns we have seen have the objective of hiding complexity from the user to avoid overwhelming them. However, this pattern seems to break this rule at first sight. When the complexity of navigation would make the features hard to find (for example, more than three levels of depth: group -> subgroup -> sub-subgroup), an alternative is to use a **mega menu**. This pattern is commonly used in government websites and other complex organizations with heavy unavoidable content. The basic concept is to present a large menu containing all (or most) of the options to choose from. This doesn't mean that there will not be a "drilling down" after reaching these sections, but it does make access faster. For example, let's see the official website for the city of Hyvinkää, Finland (www. hyvinkaa.fi):

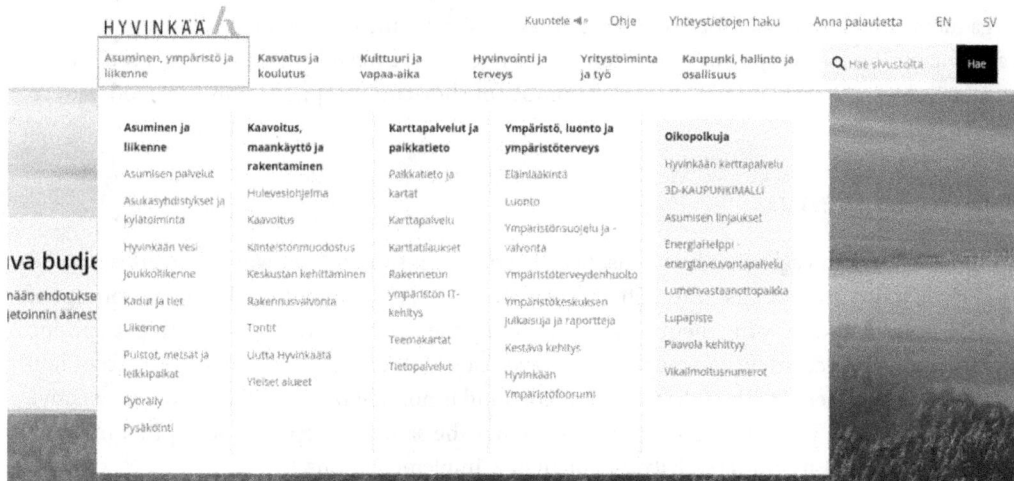

Figure11.21: Hyvinkää city home page mega menu

As you can see, it is a compelling number of options with just one click in the top menu. However, notice how well organized they are and visually separated from each other. The mega menu pattern breaks the *simplicity* rule but does not forget other rules and principles of UI design, making it still pleasing to the eye. In other cases, it is also possible to consider the mega menu as a placeholder for other patterns, such as a sidebar or accordion menu. The official Packt website (https://www. packtpub.com/) uses this pattern in the main menu **BOOKS & VIDEOS** option:

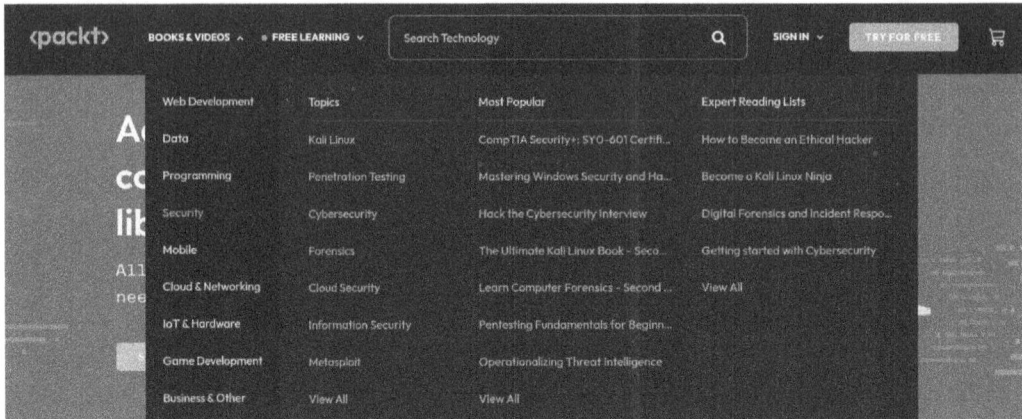

Figure 11.22 – Packt mega menu with a sidebar to categorize and filter options

A mega menu could be a place for innovation as it is easier for the user to understand and use. It is not a viable option for every web application but can be very powerful when properly used. When there is no other option than to show a large number of shortcuts or options, this is a good pattern to keep in mind.

Drop-down menus

Drop-down menus and selection boxes display a list of options to the user when they are activated (by a click or touch, for example) through a trigger area (icon, text, or button). Here, the concept of hiding the options and showing them upon "request" by the user is more important than a specific implementation. The accordion and mega menus make use of this concept, for example. The accordion menu could be considered a list of drop-down menus and is not that far off. HTML provides a native solution for selection lists (the `select` element), but the same concept can be applied in many circumstances and, with some creativity, can birth new implementations.

Common patterns for content organization

The patterns we will see next have to do with the overall organization and layout of the website or application.

Responsive applications

The term *responsive* relates to the way the layout of an application adapts to the size of the screen. Perhaps you have heard of the "mobile-first approach," which is a philosophy to design first for smaller screens, and then move upward to the possible resolutions to reach the desktop, which is considered the largest. While there are ways to accomplish this with JavaScript, the most sensible direct approach is to use well-thought-out designs with CSS media queries.

Depending on the application, there are certain formulas to create a responsive site, but analyzing the many alternatives goes beyond the objectives of this book. Instead, we will see only one as an example, using the "switch columns" method. This method basically sets for mobile (or narrow portrait screens) the main content inside a single vertical column, section by section. A main navigation bar or menu is placed at the top or bottom of the screen, always visible. Then, for the desktop, the navigation bar totally or partially moves to the top menu or to a sidebar, and the content from the main column moves to horizontal sections, stacked one after the other. This method is illustrated and better understood with this figure:

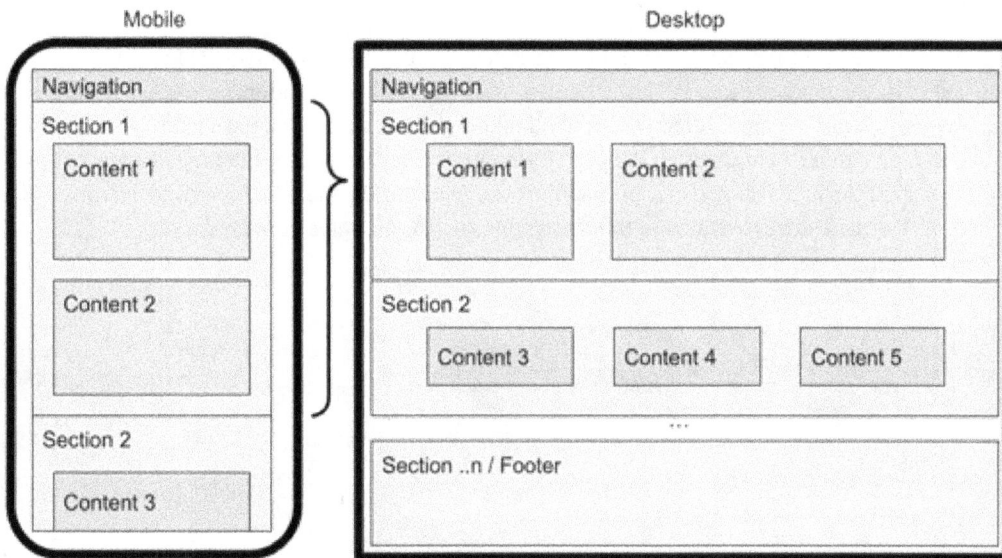

Figure 11.23 – Transformation from mobile first to desktop

As you can see from the preceding figure, the sections are always in the same order, but the content inside adapts from a vertical to horizontal layout direction. This concept is so simple and clean to design that it has become the standard for most landing pages. Once you understand this pattern, you begin to see it everywhere it is implemented.

A simple coding approach to do this change is by using the CSS flexbox model and changing the orientation from column to horizontal at the section level. Here is an example:

```
// For mobile
    @media only screen and (max-width: 600px) {
        section{
            display: flex;
            flex-direction: column;
        }
    }
```

```
// For desktop
section{
    display: flex;
    flex-direction: row;
}
```

Notice how the code has included a breakpoint at 600-px wide. You can control different screen sizes by applying multiple media queries with breakpoints.

Home link

This pattern is so ubiquitous that we do not even think about it. The main corporate logo is placed at the top left of the page, as a link to the home screen. The position is not random, and it has its origins in the way users "scan" a page. Different users will take a quick look when a page loads, guiding the eyes in a *Z*, *L*, or *T* movement through the page. Placing the logo as a link at the top left ensures that it is the first item the user will register. In this chapter, *Figures 11.21* and *11.22* are good examples of this pattern. But there are exceptions to the placement, such as Google's home page:

Figure 11.24 – Google's home page is an exception to this pattern as the logo is not a link

However, the exception shown in the preceding figure is only temporal, as Google goes back to this pattern once it presents the search results:

Figure 11.25 – Google's search result applies the pattern

The home link pattern should bring you back to either the home page or the first step of the process. This pattern is so widespread and understood by users, that any exception should be made very carefully with a good understanding of the behavior and interaction of the user base.

Hero section, call to action, and social media icons

The **hero section** is the first part that displays when the page loads in the browser and goes from the very top (where the home link and main menu reside) to, at most, the visible bottom of the screen. All content below this section is referred to as "below the fold," meaning that to see it, the user needs to scroll down the page. The hero section is considered the most important part of the home page, and the place where the initial **call to action** is placed. We visited this concept previously in this chapter when talking about UI design principles and contrast. Nowadays, most websites will adhere to this pattern and display the hero section in contrast, with large images or a carousel, and a predominant call to action.

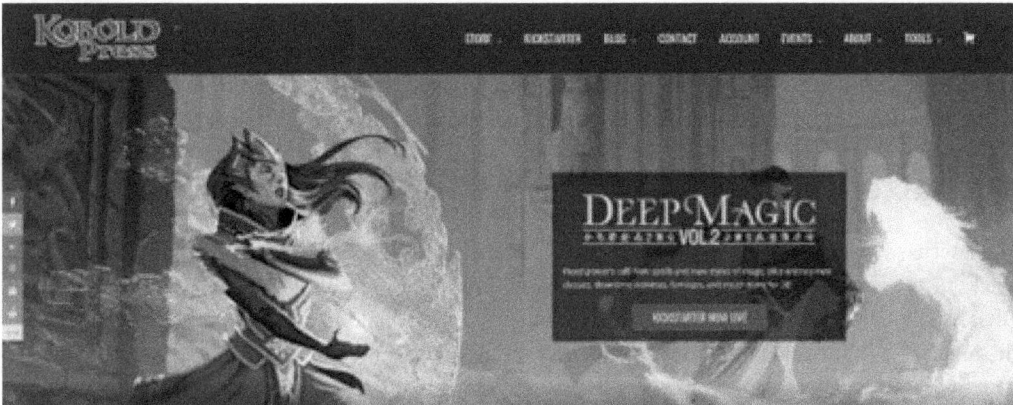

Figure 11.26 – Kobold Press home page's hero section – colorful and direct

In the preceding example from Kobold Press (`https://koboldpress.com/`), we can see how they have applied several patterns seen here, from the home link and main menu to the hero section and the call to action. Notice also, on the left side, the display of social media icons as a floating sidebar. It is becoming a pattern to place such icons with links to the social media address for each institution, or in the case of posts or articles, a link to "share" the content. Common places to locate the social icon's bar are as follows:

- In the footer of the website
- At the beginning and/or end of an article
- In the description of a product

When including links to share direct content from a Vue application (for example, a product from an online store), it is important that the link and the Vue application point directly to such an item. Careful attention needs to be placed on the way the navigation URL is formed and interpreted on the first load by the Vue application, to avoid sharing links that will open the home screen instead of the desired section of the application.

Other patterns

There are other patterns, hundreds if not thousands, that have a direct influence on the user experience. I encourage you to continue investigating these, such as the following:

- Shopping cart patterns
- User onboarding
- Gamification patterns
- Registration and de-registration patterns

There is, however, a dark side to user experience, and it has to do with the use of manipulative and deceitful practices. As an ethical developer, you should avoid using any of these dark patterns.

Dark patterns

Dark patterns are carefully designed interfaces and interactions with the sole purpose of manipulating or tricking the user into performing unintended actions or even entangling them into malicious results. After such a definition, you may think that such actions belong to the deepest shady corners of the internet. Sadly, even mainstream companies often follow these unethical practices. As a matter of fact, the reproduced examples in these sections all belong to such a category, and as often happens with design patterns, many of these overlaps or can be found nested within each other. Let's see them one by one.

Trick questions

This pattern is a simple or complex play on words to trick the user to do the opposite of what their intention is. Here is an example:

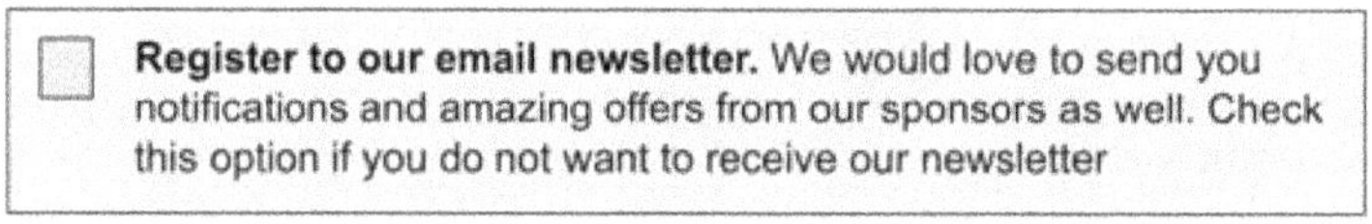

☐ **Register to our email newsletter.** We would love to send you notifications and amazing offers from our sponsors as well. Check this option if you do not want to receive our newsletter

Figure 11.27 – A trick question for a newsletter subscription

As you can see in this example, a user filling in a form would be tempted to leave this checkbox unchecked if they don't want to receive a newsletter from the company. The common tendency is to pay attention to the marked text in bold, which is a sort of title. The real action for the checkbox can be found at the end of the paragraph, which the majority of users would not read: **Check this option if you do not want to receive our newsletter.**

Sneak into the basket

This dark pattern appears in many shopping applications, quite commonly in services such as hosting and airline tickets. It consists of placing items into the shopping cart at checkout, but that the user has not selected, usually presented as an offer or a "necessary" item of a small amount. Here is an example:

Checkout	
YourAwesomeDomain.com	USD 10,00
First time setup	USD 0,50

Figure 11.28 – An additional item has been added to the shopping cart

In this example, after ordering for a new domain, a new item, **First time setup**, has been automatically added to the shopping cart. This item does not have any other explanation, and the amount appears to be "small" in comparison to the main purpose of the purchase. Often, these types of items are "scammish" and have no other intent than to add penny by penny to the final amount. In some cases, there may be options to remove such items before confirming the purchase, but quite often there are none.

Roach motels

This pattern appears quite often for services and subscriptions. It consists of making a very easy purchase, often after a free trial period, with the condition that charges will continue until explicitly canceled by the user. Here is where the dark pattern appears: by making this "unsubscribe" process complicated or impossible to complete. For example, some companies require that the user contact the support team with a signed letter accompanied by a legal ID. The basic concept is to "trap" the user so rescinding the contracted service becomes near impossible.

Privacy Zuckering

This dark pattern is named after the founder of a recognizable social media company. It consists of offering to the user a large number of services, for free, while the use of the applications monitors the user's activity and behavior. This data is then collected and sold behind the scenes to third-party companies without the knowledge of the user. Often, this practice is somehow named in a rather lengthy contract of terms of service, which the user needs to accept before using the services. This way, the company alleges that the user has given their consent and is aware, while very few users ever read or interpret properly such an agreement.

Preventing price comparison

In this pattern, the website presents to the user a number of plans for services, but purposely hides or disguises either features or individual prices, so the user is unable to make a direct comparison to select the most suitable option.

The price is hidden or disguised in such a way that the user cannot make an informed decision and must select an option based on features or other attributes.

Misdirection

This is another pattern often used in shopping cart workflows. It consists of using names and options that are confusing to the user, with pre-selected options that hide alternatives and better price deals. If the system has a floating price value (for example, for hotels or airplane tickets), this pattern is often used to manipulate the user into selecting the option that is in the best interest of the company.

Hidden costs

In this pattern, the user's selection of products or services does not disclose a total or comprehensive description of the associated costs involved (beyond taxes). Either in the initial purchase or subsequent purchases, the total amount paid results in a higher price than what the user estimated in the first instance.

Bait and switch

This pattern is widely used by online advertisers and is one that causes the most hate in captive users. It simply disguises one option to perform another or the exact opposite the user intended. A classic example is when a pop-up window displays a **Close** button (usually a simple **X**), but when the user clicks it intending to close the dialog window, it opens a new tab with the advertised website.

Confirm shaming

A highly manipulative pattern, it involves deliberately using wording or actions to shame and ridicule the user into doing something they didn't want at the beginning of the transaction. It can go from mildly annoying to outspoken insulting. It is often used together with other dark patterns. Here is an example:

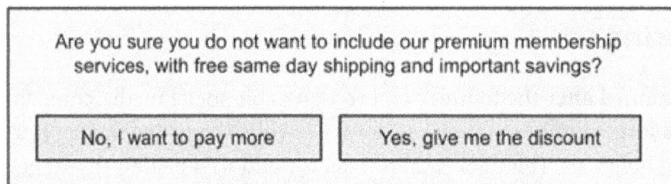

Figure 11.29 – A confirmation dialog when removing a service from the cart

Notice in the example how the wording in the action buttons is highly manipulative, even though the original question is not so. This pattern is a double-edged sword, as users may feel the rejection of the entire service and cancel the entire operation.

Disguised ads

When an advertisement is injected into a page, it can disguise itself as proper content, imitating styles and action buttons with the intent of tricking the user to trigger a redirection or download a file. In some cases, the camouflage is such that it is not possible to distinguish the site's call to action and the advertisements. This pattern is popular on free sites that offer to host files for download, where it is common to find multiple **Download** buttons on the page, yet only one does actually download the desired file and the others redirect the user to a third-party site. Here is one example:

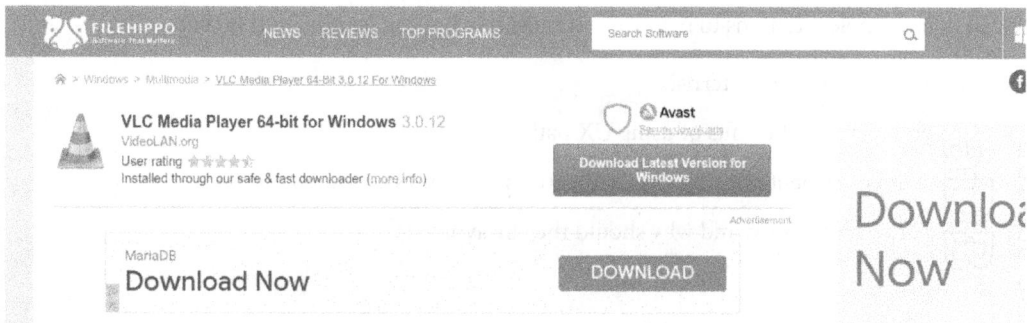

Figure 11.30 – FileHippo.com hosts free software. Some ads disguise as Download buttons

In this screenshot from `FileHippo.com`, if you access the download page for VLC Media Player, the site presents multiple **Download** buttons. If a user does not pay attention, it may trigger a different action than the intended software download.

Friendly spam

In this dark pattern, the application requests access to the user's contacts, with the idea of expanding their network or social circle. If the user accepts, their entire contact list will be "spammed" with emails as if coming from the user, offering the service. Often, once this information has been shared, it is also shared with third-party entities and advertisers.

The list of dark patterns may not be directly associated with a single media or framework, but it is the ethical responsibility of people in this industry to avoid or prevent them.

Summary

In this chapter, we saw important patterns to present the user with a satisfactory user experience. We also learned important terms to speak with designers in equal terms and with mutual understanding – a necessary point for collaboration and presenting the user base with the best possible positive experience. We also reviewed some of the most common dark patterns, which are techniques and workflow implementations to manipulate users and often deprive them of privacy and resources. While mostly informative, this chapter should give you a better understanding of the environment where web applications are built, and what standards to follow for easy use. This is all knowledge that a good engineer and developer should be aware of.

Review questions

Here are some simple questions to help you summarize what you learned in this chapter:

- What are UI and UX patterns?
- Can you name the benefits of using UX patterns?
- How can you benefit of using standard patterns in your Vue 3 components?
- What are dark patterns and why should they be avoided?

Appendix: Migrating from Vue 2

Migrating a Vue 2 application to Vue 3 is not as straightforward as just replacing the framework. While the Options API remains fully compatible and there should not be a need to migrate to the Composition API, there are other breaking changes that we should be aware of.

The changes between versions affect not only the core framework but also the ecosystem (new router, state management, etc.) and other dependencies. In Vue 3, there is also a new official bundler, **Vite** (which replaces **WebPack**), a new router and state management (**Pinia**, the replacement for **Vuex**), and other plugins as well. The list of changes included here is a quick reference to help you migrate your application but may not be exhaustive with all the nuances of specific needs for each particular project. Because of this, I will refer you to the official documentation for migration at `https://v3-migration.vuejs.org/`.

Here is a non-exclusive list of the major changes, other than the new Composition API:

- A different way to bootstrap and start the application
- Changes in global components and plugins registration
- Changes in the `data` property
- Changes to `v-model`, `props`, and `emits`
- Reactivity options
- Framework browser compatibility
- Changes in directory and file organization
- Changes in the router and state management

This list doesn't show all the changes *under the hood* that the framework went through, but it will give you a starting point to migrate your working application from Vue 2 to Vue 3. Now, let's see each of these in more detail.

A different way to bootstrap and start the application

The way to bootstrap and start our application has changed. It is now required that we import a constructor from the Vue bundle. Let's compare both implementations from `main.js`:

Vue 2 application instantiation:

```
import Vue from "vue"
const app=new Vue({el:"#app"})
```

In Vue 2, we import the Vue constructor and pass an object with options. In Vue 3, after the application has been created, we attach components, plugins, and so on before mounting our application to the top-level component. Here is the example rewritten for Vue 3:

Vue 3 application instantiation:

```
import {createApp} from "vue"
const app=createApp({..})
app.mount("#app")
```

The location of the index.html file has also changed and is now placed at the root of our application. You can see more changes to the document structure in *Chapter 3, Setting Up a Working Project*.

Register global components, plugins, and so on

In Vue 2, we declare an application-wide component (global) by attaching it to the Vue root instance. Here is an example:

```
import Vue from "vue"
import MyComponent from "MyComponent.vue"
vue.component("myComponent", MyComponent)
const app=new Vue({...})
```

In Vue 3, we instead register components and plugins with the application *after* it has been created and *before* it is mounted. The component (for components), use (for plugins), and directive (for directives) methods are all chainable. Here is how the preceding example looks in Vue 3:

```
import { createApp }from "vue"
import MyComponent from "MyComponent.vue"
const App=createApp({...})
App.component("myComponent", MyComponent).mount("#app")
```

If we do not need to reference the application, we can just concatenate the instantiation of the application as in this example:

```
import { createApp }from "vue"
import MyComponent from "MyComponent.vue"
createApp({...}).component("myComponent", MyComponent) .mount("#app")
```

The application bootstrap is independent of the syntax used to describe components (Options API, Composition API, or script setup).

The data property is now always a function

In Vue 2 applications, there is a discrepancy in the `data` attribute. The root component has a property that is directly a reactive definition, while all other components need to provide a function that returns an object as the `data` property. This created an inconsistency in the creation of components. This issue has been resolved in Vue 3, so now *all components are treated equally*, meaning the data attribute always has to be a function that returns an object whose members will be reactive properties.

Here is an example of the root component in `main.js`:

```
createApp({
    data(){return {...}}
})
```

And then in all other components, you have the following:

```
export default {
    data(){return {...}}
}
```

Notice that for these examples, we are using the Options API for clarity. When using the `script setup` syntax, you do not need to declare a `data` attribute.

There are more reactive options to choose from

When using the Composition API, we have two options to create reactive properties: `ref()` and `reactive()`. The first one returns an object with a `.value` property that is reactive. The second converts an object passed as an argument and returns the same object with reactive properties. Here is one example:

```
<script setup>
import {reactive, ref} from "vue"
const
    data=reactive({name:"John", surname:"Doe"}),
    person=ref({name: "Jane", surname:"Doe"})
    // Then, to access the values in JavaScript
    // Reactive object
    data.name="Mary"
    data.surname="Sue"
    // Reactive ref
    person.value.name="Emma"
    person.value.surname="Smith"
</script>
<template>
    <strong>{{data.surname}}, {{data.name}}</strong><br>
    <strong>{{person.surname}}, {{person.name}}</strong>
</template>
```

Notice the difference in syntax. At this point, you may think about when to use one or the other. Here is a small comparison of when to use each one:

`ref()`	`reactive()`
<ul><li>Applies to any data type, not only primitives.</li><li>When applied to objects or arrays, you can replace them.</li><li>It uses getters and setters to detect changes and trigger reactivity.</li><li>Use it by default for simple data. For arrays and objects (complex types), it is recommended to use `reactive()` when working with their internal elements. When the entire object will be replaced, it is convenient to use `ref()`.</li></ul>	<ul><li>Applies to objects and arrays, but not primitives. Makes their attributes reactive.</li><li>The object cannot be replaced, only its attributes.</li><li>It uses the native implementation of the `Proxy()` handlers to detect changes and trigger reactivity.</li><li>Use when you need to group a large number of variables that must "travel" together.</li></ul>

Table A.1 - A simple guide to choose between ref() and reactive()

Each method has its own advantages. From the point of view of reactive properties with complex types, it doesn't matter which one you use. In some cases, `reactive()` can be more performant due to the use of native implementations in the browser.

Changes to v-model, props, and events

This is a big change from Vue 2 that can and will break your code. In Vue 3, we no longer receive and emit the property value. Instead, any prop can be input/output, such as `v-model`. The default `v-model` attribute is received in a *prop* named `modelValue`, and the counterpart *emit* prepends `update:`, so it is called `update:modelValue`.

In Vue 3, we can now have multiple *v-models* at the same time. For example, we can have `v-model:person="person"` in our component, and define the prop as `"modelPerson"` and the event as `"update:modelPerson"`.

Props and emits are now macros (a macro is a special function provided by the bundler or framework). **Props** have the same footprint as in Vue 2, so you can define them as arrays, objects, include types, default values, and so on.

Here is an example with a default v-model and a notated model:

```
const $props=defineProps(['modelValue','modelPerson']),
    $emit=defineEmits(['update:modelValue','update:modelPerson'])
```

Props and emits are discussed in this book in more detail in *Chapter 4, User Interface Composition with Components*.

Removed old browser compatibility

Vue 3 was built for speed and "modern" JavaScript. Backward compatibility for older browsers has been removed. Many internal functions used for reactivity now use native implementations by default (for example, the Proxy API). If you need to support an application in an outdated browser, you should consider staying with Vue 2, but fear not! There is an official plugin for Vue 2 to use the new *Composition API*, including the `script setup` syntax:

- Vue 2.7 includes it without plugins (`https://blog.vuejs.org/posts/vue-2-7-naruto.html`)

- If you are on Vue 2.6 or below, you can find the plugin here: `https://github.com/vuejs/composition-api`

- If you still want the speed of Vue 3, there is a special migration build that has almost the same API as Vue 2 (see `https://v3-migration.vuejs.org/migration-build.html`)

- Why remove old browsers' compatibility? There are many reasons, including the following:

  - The global usage of older browsers has fallen below a significant percentage, and it is expected to continue to drop over time.

  - With the removal of old code and compatibility checks, the resulting Vue core implementation is lighter and more performant. The increase in speed and reduction in bundle size is significant, making our applications load faster and be more responsive.

In practice, there are two browser engines that take up most of the market: browsers based on Chromium, and ones based on Mozilla Firefox. Check `www.caniuse.com` if you need to use a feature that may not be available in older browsers.

Changes in directory and file organization

The organization for the directory structure in Vue 2 was influenced to some degree by the bundler at the time, **WebPack**, and the **Vue CLI**. Now that Vue 3 has moved to **Vite**, files have been organized to better reflect the development workflow. Such is the case of `index.html`, which has moved to the root folder out of the `Public/` folder. It now has a more prominent place in the bundling process. This and other changes appear in *Chapter 3, Setting Up a Working Project*.

Changes in the router and state management

The new approach to components and modularity also affects the router and the state management. While a new version of the router has been provided for Vue 3, the state management's official solution has moved away from **Vuex** to **Pinia**. More information about the new router and Pinia can be found in *Chapter 5*, *Single-Page Applications*, and in *Chapter 7*, *Data Flow Management*, respectively.

The new router now has a different approach to defining modes, using constructors such as `createWebHashHistory` (hash mode), `createWebHistory` (history mode), and `createMemoryHistory` (navigation in memory alone). This change also affected the configuration of the production bundle. In WebPack, when in history mode, the deployment path was part of the bundler configuration. Now, the path is passed to the constructor as a parameter, being handled completely by the router.

New components and other changes

Vue 3 also introduces new components such as `teleport` (a special component that allows placing reactive templates outside the Vue component tree, inside another DOM element), but also breaks free of some limitations in Vue 2. For example, components now can have more than one root element. Please refer to the official documentation to learn more about the new components in Vue 3.

Other breaking changes

To see a full list of breaking changes not mentioned here, please check the official documentation at `https://v3-migration.vuejs.org/breaking-changes/`.

Summary

Migrating from Vue 2 to Vue 3 has a clear path, with only a few breaking changes to be aware of. The new Composition API however, does require a change of mentality, but it comes naturally when using the `script setup` syntax. But the most important feature of Vue 3 is the performance gains and the size reduction. In short, Vue 3 is fast, very fast, and well worth the migration. For projects supporting outdated browsers, there are plugins for the Vue 2.x branch that provide some of the advantages of Vue 3, but for other projects seeking the positive gains of Vue 3, it is well worth the migration.

Final words

Congratulations on reaching the end of this book! We have covered a wide spectrum of topics, from the very basics of Vue to the deployment of the final product. Let's review together the main concepts for each chapter:

- In *Chapter 1, The Vue 3 Framework*, we introduced the key Vue concepts and the different syntax options available for writing components

- In *Chapter 2, Software Design Principles and Patterns*, we took a deep-dive into important conceptual and well-tested patterns for architecting our code

- In *Chapter 3, Setting Up a Working Project*, and *Chapter 4, User Interface Composition with Components*, we learned how to start a Vue project and how to translate designs into working code

- In *Chapter 5, Single-Page Applications*, and *Chapter 6, Progressive Web Applications*, possibly the most important chapters, we learned how to create advanced applications with navigation and installation via the browser's native functions

- In *Chapter 7, Data Flow Management*, and *Chapter 8, Multithreading with Web Workers*, we learned more about how to improve performance and control the information flow with good practices

- *Chapter 9, Testing and Source Control*, introduced tools to automate programmatic tests to secure the good quality of our code

- *Chapter 10, Deploying Your Application*, gave us a view into the steps and resources needed to publish and secure our server with a secure protocol

- *Chapter 11, Bonus Chapter -UX*, gave us a view from the user perspective, and a common vocabulary to collaborate with UI/UX designers

Indeed, this has been a long journey, but I'm confident and positive that this content will improve your skills as a developer and professional.

Where to go from here

Technology keeps advancing daily, so there is much to learn ahead. New tools and patterns are created regularly. Because of this, it is not possible to cover them all in just one book. Often, while preparing a chapter, I've been limited to touching on some technologies and concepts only at a surface level due to the scope and breadth of this book. For example, beyond the web, Vue can also be used to develop hybrid applications with tools such as **NW.js** (`https://nwjs.io/`), **Electron** (`https://www.electronjs.org/`), **Tauri** (`https://tauri.app/`), and more.

Learning about this framework and the technologies it is based upon will give you important skills.

Finally...

I express my appreciation for your dedication to this discipline and thank you for purchasing this book. I wish you well and brilliant success in your future endeavors and professional career.

Sincerely,

Pablo David Garaguso

www.pdgaraguso.com

Index

A

B

V

Visual Studio Code 52
  URL 52
Vite 6, 7, 53, 77, 251, 255
  configuration options 60, 61
  plugins 61
Vite-PWA plugin 138-140
Vitest 187, 189
  asynchronous code 192
  benefits 189- 191
  fail on purpose 192
  installing 189
  UI 198, 199
  URL 189
v-model directive 80, 81
Vue 3 1, 43
  built-in components 12, 13
  built-in directives 10-12
  reactivity 3
  using, in web application 4, 5
Vue 2 application
  bootstraping and starting, ways 251
  data property 253
  events, modifying 254
  global components, registering 252
  modifying, in directory and
    file organization 255
  props, modifying 254
  reactive options, selecting 253, 254
  v-model, modifying 254
Vue 3 application
  modifying, in router and state
    management 255
  old browser compatibility, removing 255
  upgrading 256

Vue 3 router 99
  installation 99, 100
  named views 106
  nested routes 106
  nested routes' definition 107-111
  programmatic navigation 106
  router object 101-103
  routes' definition 101-103
  template components 103-105
Vue CLI 255
Vue-material 57
Vue project
  folder structure 55, 56
  integration, with CSS frameworks 57, 58
  setting up, with Vite 53, 54
Vue Test Utils 187, 193
  installing 193-195
  URL 193
Vuex 251

W

w3.css framework 58
  URL 58
we3js
  reference link 123
Web 1.0 98
Web 2.0 98
Web 3.0 98
Web3 authentication 122
web application
  considerations, for publishing 209
  protecting, with Let's Encrypt 217, 218
  Vue, using in 4, 5
Web history 102
WebPack 251, 255

‹packt›

Subscribe to our online digital library for full access to over 7,000 books and videos, as well as industry leading tools to help you plan your personal development and advance your career. For more information, please visit our website.

Why subscribe?

- Spend less time learning and more time coding with practical eBooks and Videos from over 4,000 industry professionals

- Improve your learning with Skill Plans built especially for you

- Get a free eBook or video every month

- Fully searchable for easy access to vital information

- Copy and paste, print, and bookmark content

Did you know that Packt offers eBook versions of every book published, with PDF and ePub files available? You can upgrade to the eBook version at packtpub.com and as a print book customer, you are entitled to a discount on the eBook copy. Get in touch with us at customercare@packtpub.com for more details.

At www.packtpub.com, you can also read a collection of free technical articles, sign up for a range of free newsletters, and receive exclusive discounts and offers on Packt books and eBooks.

Other Books You May Enjoy

If you enjoyed this book, you may be interested in these other books by Packt:

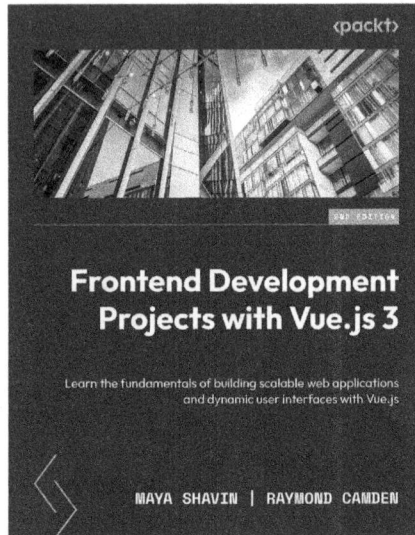

Frontend Development Projects with Vue.js 3 - Second Edition

Maya Shavin, Raymond Camden

ISBN: 9781803234991

- Set up a development environment and start your first Vue.js 3 project
- Modularize a Vue application using component hierarchies
- Use external JavaScript libraries to create animations
- Share state between components and use Pinia for state management
- Work with APIs using Pinia and Axios to fetch remote data
- Validate functionality with unit testing and end-to-end testing
- Get to grips with web app deployment

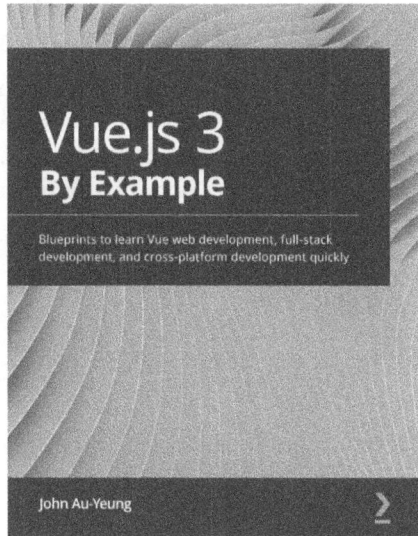

Vue.js 3 By Example

John Au-Yeung

ISBN: 9781838826345

- Get to grips with Vue architecture, components, props, directives, mixins, and other advanced features
- Understand the Vue 3 template system and use directives
- Use third-party libraries such as Vue Router for routing and Vuex for state management
- Create GraphQL APIs to power your Vue 3 web apps
- Build cross-platform Vue 3 apps with Electron and Ionic
- Make your Vue 3 apps more captivating with PrimeVue
- Build real-time communication apps with Vue 3 as the frontend and Laravel

Packt is searching for authors like you

If you're interested in becoming an author for Packt, please visit `authors.packtpub.com` and apply today. We have worked with thousands of developers and tech professionals, just like you, to help them share their insight with the global tech community. You can make a general application, apply for a specific hot topic that we are recruiting an author for, or submit your own idea.

Hi!

I am Pablo D. Garaguso, author of *Vue.js 3 Design Patterns and Best Practices*. I really hope you enjoyed reading this book and found it useful for increasing your knowledge and understanding of the amazing Vue 3 framework and the many possibilities to write first class applications.

It would really help me (and other potential readers!) if you could leave a review on Amazon sharing your thoughts on *Vue.js 3 Design Patterns and Best Practices*.

Go to the link below or scan the QR code to leave your review:

https://packt.link/r/1803238070

Your review will help me to understand what's worked well in this book, and what could be improved upon for future editions, so it really is appreciated.

Best Wishes,

Download a free PDF copy of this book

Thanks for purchasing this book!

Do you like to read on the go but are unable to carry your print books everywhere?

Is your eBook purchase not compatible with the device of your choice?

Don't worry, now with every Packt book you get a DRM-free PDF version of that book at no cost.

Read anywhere, any place, on any device. Search, copy, and paste code from your favorite technical books directly into your application.

The perks don't stop there, you can get exclusive access to discounts, newsletters, and great free content in your inbox daily

Follow these simple steps to get the benefits:

1. Scan the QR code or visit the link below

`https://packt.link/free-ebook/9781803238074`

2. Submit your proof of purchase
3. That's it! We'll send your free PDF and other benefits to your email directly

www.ingramcontent.com/pod-product-compliance
Lightning Source LLC
Chambersburg PA
CBHW080517220326
41599CB00032B/6118